AIR, WATER AND SOIL POLLUTION SCIENCE AND TECHNOLOGY

AIR POLLUTION

MANAGEMENT STRATEGIES, ENVIRONMENTAL IMPACT AND HEALTH RISKS

Air, Water and Soil Pollution Science and Technology

Additional books in this series can be found on Nova's website under the Series tab.

Additional e-books in this series can be found on Nova's website under the eBooks tab.

AIR, WATER AND SOIL POLLUTION SCIENCE AND TECHNOLOGY

AIR POLLUTION

MANAGEMENT STRATEGIES, ENVIRONMENTAL IMPACT AND HEALTH RISKS

GERALD L. BURNS
EDITOR

nova publishers
New York

Library of Congress Cataloging-in-Publication Data

ISBN: 978-1-63485-374-3

Published by Nova Science Publishers, Inc. † New York

CONTENTS

PREFACE

This book provides new research on the management strategies, environmental impacts and health risks of air pollution. Chapter One analyzes historical air pollution data via linear regression and the time-series technique. Chapter Two provides a spatial analysis of air pollution from road transport within urban areas and its relation to health risks. Chapter Three presents the findings on the effects of short- and long-term exposure to air pollution on death rates in Belgrade, Serbia, based on a 6-year regular pollutant monitoring and the corresponding administrative records on mortality. Chapter Four discusses how bioaerosols impact the environment, and provides methods for reducing health risks. Chapter Five reviews a case study on heavy metals air pollution of mines in the Bregalnica River Basin.

Chapter 1 – Pollution data in eleven categories obtained over the 1980 – 1989 period in four highly industrialized nations (USA, UK, West Germany, The Netherlands)are examined in terms of linear regression and the time-series technique, treating time and annual population size, respectively, as explanatory variables, and the levels of polluting substances as the dependent variable. The pollution level versus time relationship can be adequately described by simple univariate linear regressions in slightly more than one half of the pollutant categories presented in the ten – year compilation of data, with acceptably high coefficients of determination (R^2). Removal of autocorrelation between adjacent residuals, based on the Durbin – Watson statistic and the concept of the lag-one autocorrelation coefficient, is illustrated in the case of two specific kinds of emission. The results underline the importance of a fully quantitative understanding of all ecologically meaningful explanatory variables in order to fully utilize the potential of the analysis in combating air pollution via preventive measures based on pertinent numerical cause – effect relationships.

Chapter 2 – This study extends the traditional research in air pollution and health risks by using spatial and temporal methods in geographical information system (GIS). The spatial interpolations are produced from point samples, which are represented by sites of monitoring stations, and by randomly generated sites with local estimates of urban air pollution by geographically weighted regression (GWR). The explanatory variables for estimates of urban air pollution, NO_2 and PM_{10}, are represented by the elevation, the distance to the nearest major road, and the ratio of the built-up area in the defined circles. The dependent variables contain annual average concentration of NO_2 and PM_{10} in Prague. The attached residuals and multiple R-squared indicate that each GWR model can partially explain the variation of concentrations in relation to the selected exploratory variables. Studying the relations between urban air pollution, such as NO_2 and PM_{10}, and lung cancer incidence are provided separately for ten districts of the city, independently for males and females. In spite of that the spatial analysis of urban air pollution on a reduced number of monitoring stations and limited available spatial datasets can be beneficial for decision-making processes, relations of the lung cancer incidence to urban air pollution indicate weak or even non-existed associations. The available datasets about the lung cancer incidence are mostly based on aggregated information for each district in the city. Thus, there is a little evidence about original sites of the lung cancer incidence, which documents many studies that are focused on assessing the public health impacts by urban air pollution. Presented spatial and temporal analysis of air pollution from road-transport within the urban areas represents a new way of research that is based on integration of urban air pollution monitoring, understanding potential human exposure, and advanced modeling tools implemented in ArcGIS. It can be used for risk assessment, decision-making processes and urban planning. It is anticipated that the use of advanced methods of exploratory spatial data analysis will bring new progress into the studied processes and new insights for urban planning and health protection.

Chapter 3 – Substantial scientific evidence suggests that exposure to even modest levels of air pollutants is associated with premature mortality. While relative contributions of pollutant species remain a matter of scientific debate, the current findings converge on the statement that short-term exposure over a period of few hours to weeks can aggravate the pre-existing health conditions, sometimes leading to a fatal outcome, whereas longer-term exposure reduces life expectancy by a few years. Susceptible groups including children and the elderly are at greater risk, as well as the people living in overpopulated low-income regions, due to being exposed to pollutant concentrations up to twenty times higher than those commonly observed in modern countries. The goal of this study is to present the findings on the effects of short- and long-term exposure to air pollution on death rates in Belgrade area (Serbia), based on the 6-year (2009-2014) regular pollutant monitoring and the corresponding administrative records on mortality. To that end, a quasi-Poisson regression model was combined with distributed lag non-linear model to examine this association after controlling for confounding variables such as temperature, seasonal trends and day of the week. The study discusses relative rates, expressed as % increase in circulatory, respiratory and malignant mortality per unit increase in daily average pollutant concentrations, together with the observed age- and gender-specific effects. According to the results, short-term exposure to elevated concentrations of PM_{10}, SO_2 and NO_2 is associated with relatively small mortality risk, whereas exposure during the last year appears to be more significant, particularly for circulatory mortality. Apart from the elderly group, severity of health effects is also observed for the male population under the age of 65. According to multiple economic indicators, Serbia is considered a developing country, even though its demographic trends are complementary with or less favourable than those typically anticipated in Western developed countries, notably due to high circulatory and malignant death rates. Poor air quality in Belgrade area is identified as an important environmental risk factor with pollutant concentrations exceeding the values observed in most European metropolis.

Chapter 4 – Bioaerosols (biological particles derived from viruses, bacteria, fungi, plants and animals) are ubiquitous in the air. This organic dust is considered to be hazardous to human health. Acute and chronic diseases are health risks associated with viable and non-viable airborne biological particles because they can be toxic, allergenic and infectious. With increasing awareness of the environmental impact of bioaerosols, which can spread easily with air currents over long distances, the development of efficient air filtration system is important for preventing health risks. Air filtration system is the most used method for controlling air pollution with bioaerosols but it is not always efficient because microbes can survive and proliferate on the surface of filters and be potential sources of diseases and allergies. Lately, many researchers have focused on improving indoor air quality by removing and inactivating biological agents. Successful methods for control and inactivation of bioaerosols include filters coated with antimicrobial nanoparticles from different plants, such as *Sophora flavescens, Euscapis japonica* or *Melaleuca alternifolia*. Filters coated with plant extracts have filtration efficiency from 77 to 99%, while the range for antimicrobial activity for these filters is from 14 to 100%. These natural products are considered ecological, environmentally friendly, cost–effective, less toxic (compared with inorganic nanomaterials) and efficient material for antimicrobial filter media development and can be easily applied to conventional air–conditioning systems.

Chapter 5 – Application of several moss species and attic dust for monitoring of anthropogenic impact on heavy metals air pollution in Bregalnica River Basin, Republic of Macedonia, was studied. Moss samples were reviewed for their potential to reflect heavy metals air pollution. The attention was focused on their quantification ability, underlying the metal accumulation within moss plant tissue and attic dust "historical archiving." Potential "hot spots" were selected in areas of copper mine (Bučim mine) and lead and zinc mines (Zletovo mine and Sasa mine) as main metal pollution sources in the Eastern part of the Republic of Macedonia. Continuously, dust distribution from ore and flotation tailings occurs. This results with air-introduction and deposition of higher contents of certain metals. Several moss species (*Hypnum cupressiforme*, *Homalothecium lutescens* and *Scleropodium purum*) were used as plant sampling media. Determination of chemical elements was conducted by using both instrumental techniques: atomic emission spectrometry with inductively coupled plasma (ICP-AES) and mass spectrometry with inductively coupled plasma (ICP-MS). Combination of multivariate techniques (PCA, FA and CA) was applied for data processing and identification of elements association with lithogenic or anthropogenic origin. Spatial distribution maps were constructed for determination and localizing of narrower areas with higher contents of certain anthropogenic elements. In this way influences of selected human activities in local (small scale) air pollution can be determined. Summarized data reveal real quantification of the elements distribution not only in order determination of hazardously elements distribution, but also present complete characterization for elements deposition in mines environs.

In: Air Pollution
Editor: Gerald L. Burns
ISBN: 978-1-63485-374-3

Chapter 1

An Analysis of Certain Historical Air Pollution Data Via Linear Regression and the Time - Series Technique

Thomas Z. Fahidy
Department of Chemical Engineering, University of Waterloo, Waterloo, Ontario, Canada

Abstract

Pollution data in eleven categories obtained over the 1980 – 1989 period in four highly industrialized nations (USA, UK, West Germany, The Netherlands)are examined in terms of linear regression and the time-series technique, treating time and annual population size, respectively, as explanatory variables, and the levels of polluting substances as the dependent variable. The pollution level versus time relationship can be adequately described by simple univariate linear regressions in slightly more than one half of the pollutant categories presented in the ten – year compilation of data, with acceptably high coefficients of determination (R^2). Removal of autocorrelation between adjacent residuals, based on the Durbin – Watson statistic and the concept of the lag-one autocorrelation coefficient, is illustrated in the case of two specific kinds of emission. The results underline the importance of a fully quantitative understanding of all ecologically meaningful explanatory variables in order to fully utilize the potential of the analysis in combating air pollution via preventive measures based on pertinent numerical cause – effect relationships.

LIST OF PRINCIPAL SYMBOLS

a, b regression parameters in a simple univariate linear regression relating Y_t to t

a_k regression parameters in a multivariate linear regression

DW Durbin – Watson statistic

e_k k – th residual (in a first – order autoregressive process)

k running index (related to time)

n dimension of a first – order autoregressive process

P_t population size at time t

Q_t manipulated population size at time t

R^2 coefficient of determination

RD residual distribution

r_1 first – order lag coefficient

t time

u_t, v_t, z_t manipulated explanatory variables

X_t explanatory variable; x_t its, or its manipulated numerical value at time t

Y_t dependent variable (emission level) at time t; y_k its value at the k-th observation time

α, β regression parameters in a simple univariate linear regressions relating Y_t to X

ξ_k random error belonging to a normal (Gaussian) distribution with zero mean and a constant variance

ρ autocorrelation parameter: the coefficient of correlation between adjacent errors

1. INTRODUCTION

Pursuant to the release of a cornerstone environmental report in the late last century [1], a comprehensive data compilation became subsequently available [2] showing year – to – year variations in the extent of air pollution in certain technologically advanced nations toward the end of the XX-th century. The aim of this chapter is to analyze some of the information provided by Kiely in eleven tables of Chapter 8 of his work, summarized in Table 1, via statistical regression and the time – series technique. Incentive for this effort stems from the intent to facilitate a proper quantitative interpretation of ecological information carried by the above - cited sources. Graphical presentation (Figures 8.3, 8.4, 8.5, 8.7, 8.8, 8.9, 8.12) of emission levels (Y) versus time (t) [2] suggests that a simple linear regression could adequately represent their relationship in most cases within the 1980 – 1989 decade, constituting the first purpose of the herein described study. The second purpose is (i) to examine the viability of the hypothesis that the variation in substance/emission levels can be interpreted as a first order time series, and (ii) to illustrate attempts to remove autocorrelation between subsequent observations in the time - series via two representative examples.

Table 1. Data Base for the Regression – and Time – Series – Based Analysis

1	8.6 [341]	Total CO per capita, kg/capita per year
2	8.7 [342]	Total CO from mobile (M) and stationary (S) sources, tonne per year
3	8.8 [343]	Total NO_2 per capita, kg/capita per year
4	8.9 [344]	NO_2 from mobile (M) and stationary (S) sources, kilotonne per year
5	8.10 [345]	Total SO_2 per cpita, kg/capita per year
6	8.11 [346]	SO_2 from mobile (M) and stationary (S) sources, kilotonnee per year
7	8.18 [350]	Total hydrocarbon per capita, kg/capita per year
8	8.19 [351]	Hydrocarbon from mobile (M) and stationary (S) sources, kilotonne per year
9	8.20 [352]	Lead from vehicles per capita, kg Pb/capita per year
10	8.14 [347]	Total particulate matter (PM), kg/capita per year
11	8.15 [348]	Particulates from mobile (M) and stationary (S) sources, kilotonne per year

Substance/emission identifier index.
SEI, Table, [page no.] [2] Substance/emission and unit.

2. THEORETICAL FOUNDATIONS AND TREATMENT OF OBSERVATIONS

2.1. Linear Regression

Amply discussed in the probability theory/statistics literature, the linear regression posited in the current study as

$$Y = a + bt \tag{1}$$

may be considered to be an admissible representation of the $(Y{:}t)$ association if (a) $R^2 \geq 0.5$, and (b) the distribution of the residuals: $e_k = y_k - y_{k-1}$; $k = 1,\ldots,n$ is essentially random, the residuals possessing zero mean and a constant variance. The latter requirement is difficult to satisfy in practice and is often ignored in computing the regression parameters $a;b$, without further investigation. The coefficient of determination, being the ratio of the variation explained by the regression to the total variation in the dependent (or response) variable, is a measure of the goodness of fit by the regression.

As shown by the R^2 – entries in Table 2, the first requirement is violated in somewhat less than one – tenth of the cases, implying that about nine – tenth of the regressions with their (a; b) values would obey Eq. (1). The second requirement, whose failure is indicated by the numeral '3,' is violated by about sixteen percent of the regressions. Moreover, approximately seventeen percent fail on account of violating *both* requirements. The simple linear regression in Eq.(1) is essentially satisfied in slightly more than one half (about 57%) of the reported cases.

2.2. Analysis via the Time – Series Technique

While descriptive data are setting up effective environmental control initiatives, year – to – year variations in the amount of polluting agents do not directly provide a numerical relationship between explanatory variables (causation/source variables) and the extent of pollution, i.e., the response variable Y. One logical explanatory variable could arguably be the population size X in the four nations of Table 2 within the 1980 – 1989 decade. A fundamental condition for the validity of the $Y_t = \alpha + \beta X_t$ regression, that the residuals be independent from observation to observation with zero mean, cannot be a - priori guaranteed, nor can the randomness of their distribution. In the case of a first – order autocorrelation relationship for the residuals:

$$e_k = \rho e_{k-1} + \xi_k; k = 1,\dots,n; |\rho| < 1 \quad (2)$$

the ξ_k parameters are routinely assumed to be independent, random and belonging to the normal (Gaussian) distribution with zero mean. Under such conditions the regression

Table 2. Fundamental characteristics of regression - based analysis of emission data [2] from the 1980– 1989 decade. R^2: coefficient of determination; RD: residual distribution 1:essentially random, 2: approximately random, 3:not random; b: slope of the linear regression, denoting the linearized annual rate of substance decrease/increase; * immaterial due to $R^2 < 0.5$

SEI	Parameter	USA	UK	West Germany	Netherlands
1	R^2	0.87	0.92	0.82	0.84
	b	- 0.012	0.0030	- 0.006	- 0.0023
	RD	1	2	1	2
2M	R^2	0.98	0.92	0.76	0.85
	b	- 1.930	0.202	- 0.236	- 0.028
	RD	1	3	2	3
2S	R^2	0.02	0.54	0.82	0.01
	b	*	- 0.016	- 0.101	*
	RD	3	2	2	3
3	R^2	0.73	0.61	0.07	≈ 0
	b	- 1.217	0.570	*	*
	RD	1	3	3	3
4M	R^2	0.81	0.90	0.91	0.48
	b	- 0.142	0.046	0.036	*
	RD	1	1	1	1
4S	R^2	0.22	0.23	0.87	≈ 0
	b	*	*	- 0.048	*
	RD	2	1	2	3
5	R^2	0.87	0.67	0.91	0.86
	b	- 2.108	- 1.921	- 3.800	- 2.097
	RD	1	2	2	1
6M	R^2	0.30	0.10	0.83	0.012
	b	*	*	- 0.0037	*
	RD	2	3	2	3
6S	R^2	0.76	0.62	0.92	0.87
	b	- 0.313	- 0.099	- 0.230	- 0.029
	RD	2	2	3	1
7	R^2	0.81	0.76	0.42	0.94
	b	- 2.483	0.236	*	- 0.903
	RD	1	3	3	3
8M	R^2	0.93	0.84	0.21	0.53
	b	- 0.172	0.017	*	- 0.002
	RD	2	2	1	1
8S	R^2	0.30	0.27	0.96	0.97
	b	*	*	- 0.019	- 0.009
	RD	2	2	1	2
9	R^2	0.94	0.79	-	0.81
	b	- 0.033	- 0.012	-	- 0.009
	RD	2	2	-	3
10	R^2	0.76	0.35	0.91	0.92
	b	- 0.842	*	- 0.339	- 0.685
	RD	2	3	1	2
11M	R^2	0.75	0.92	0.81	0.03
	b	0.017	0.009	0.001	*
	RD	3	2	3	3
11S	R^2	0.67	0.58	0.96	0.92
	b	- 0.182	- 0.011	- 0.023	- 0.009
	RD	2	2	3	2

$$Y_k = a + bx_k; k = 1,\ldots,n \tag{3}$$

represents a simple first – order autoregressive process. To test the (null) hypothesis that ρ =0, i.e., that the residuals are independent, the Durbin – Watson statistic

$$DW = \frac{\sum_{k=2}^{n}(e_k - e_{k-1})^2}{\sum_{k=1}^{n} e_k^2} \tag{4}$$

is commonly employed [e.g., 3 – 6] in the following general manner [7]. If $DW < 2$, the residuals are positively autocorrelated; if $DW \approx 2$, the residuals are uncorrelated; if $DW \approx 0$, the residuals are positively and strongly autocorrelated; if $DW > 2$, the residuals are negatively autocorrelated; if $DW \approx 4$, the residuals are negatively and strongly autocorrelated. When the observation set is large enough, tighter rules apply (Section 3.2) for determining the status of autocorrelation. The DW statistic has been predominantly employed in econometric and managerial environments, but adopting it to the domain of chemical sciences has also been proposed (e.g., [8, 9]).

2.3. Removal of Autocorrelation: An Elementary (Introductory) Approach

Once the existence of autocorrelation is ascertained, some quantitative procedures may be tested for its removal, but without prior guarantee of success,. In the Cochrane – Orcutt procedure [10] the lag-one autocorrelation coefficient [11], a statistical estimator of ρ:

$$r_1 = \frac{\sum_{k=2}^{n} e_k e_{k-1}}{\sum_{k=1}^{n} e_k^2} \tag{5}$$

is computed in the first step for obtaining the modified regression variables

$$Y_t' = Y_t - r_1 Y_{t-1} \tag{6a}$$

$$X'_t = X_t - r_1 X_{t-1} \tag{6b}$$

which are subjected to linear regression in the second step. By contrast, the procedure described by Salvatore and Reagle [12] first employs the numerical value of the a_2 coefficient in the linear regression

$$Y_t = a_0 + a_1 X_t + a_2 Y_{t-1} + a_3 X_{t-1} \tag{7}$$

instead of r_1 when applying Eqs.(6a) and (6b). In the second step the values of the first array entries are computed as

$$Y'_1 = (1 - a_2^2)^{1/2} Y_1 \text{ and } X'_1 = (1 - a_2^2)^{1/2} X_1$$

For the removal to be successful, the new *DW* value has to be near 2.

Sophisticated approaches to the subject of lag-one autocorrelation (e.g., by Riviello and Baratvas [13]) are beyond the scope of the current discussion.

2.4. Quantitative Aspects of the Time Series – Based Treatment

A logical choice of the explanatory variable is the size of population in the two countries selected for analysis, owing to the easy availability of annual variations. The first one concerns the annual total CO emissions from mobile sources in the UK during the 1980-1989 period [14], whose framework for autocorrelation removal is shown in Table 3. The regressions

$$\hat{Y}_t = -133.1822 + 2.4368 x_t \tag{8}$$

and

$$\hat{Y}_t = 3.9725 + 0.002683 u_t \tag{9}$$

with a high $R^2 \approx 0.97$, and approximately random residual distribution indicate a good fit. However, as demonstrated in the second and fifth column of Table 5, positive autocorrelation between adjacent values exists.

The annual total emissions of sulfur dioxide in the USA during the same decade constitute the second illustration in Table 4. The regressions

$$\hat{Y}_t = 98.9971 - 0.9743 v_t \tag{10}$$

and

$$\hat{Y}_t = 79.8915 + 0.9747 z_t \tag{11}$$

with a reasonably high $R^2 \approx 0.87$, and an approximately random residual distribution similarly indicate a good fit.

Table 3. Framework for autocorrelation analysis in the case SEI = 2M in Tables 1 and 2. $P_t \equiv$ UK population in thousands; $P_1 = 56284$ [14]. $Y_t \equiv$ megatonne CO

Year	Year index	$x_t=10^{-3}P_t$	$u_t=10^3(P_t-P_1)$	Y_t
1980	1	56.284	0	3.896
1981	2	56.343	59	3.938
1982	3	56.324	40	4.109
1983	4	56.301	17	4.161
1984	5	56.362	78	4.335
1985	6	56.481	197	4.431
1986	7	56.618	334	4.658
1987	8	56.743	459	5.074
1988	9	56.860	576	5.355
1989	10	56.996	712	5.792

Table 4. Framework for autocorrelation analysis in the case of SEI = 5 in Tables 1and 2. Q_t : USA population in millions on July 1 in each year rounded to the nearest ten thousand. Y_t = kg SO_2 / capita per year. Q_1 = 227.22; Q_{10} = 246.82

Year	Year index, t	Explanatory variable		Y_t
		$v_t = Q_t - Q_1$	$z_t = Q_{10} - Q_t$	
1980	1	0	19.60	103
1981	2	2.25	17.35	98
1982	3	4.44	15.16	92
1983	4	6.57	13.03	89
1984	5	8.60	11.00	91
1985	6	10.70	8.90	88
1986	7	12.97	6.69	86
1987	8	15.07	4.53	82
1988	9	17.28	2.32	83
1989	10	19.60	0	83

Pertinent aspects of dealing with autocorrelation are summarized in Table 5. The second and the fifth column indicate the existence of positive autocorrelation between adjacent residuals. Attempts to remove autocorrelation by the two procedures described in Section 2.3 and summarized in columns 3, 4, 6 and 7 demonstrate the difficulty accompanying this effort. The two instances where *DW* and its approximation [16] by

$$DW \cong 2(1-r_1) \tag{12}$$

are reasonably close to the numerical value of 2 suggest at most a feeble remaining autocorrelation.

Table 5. Removal of autocorrelation by the Cochrane – Ormutt (A) and the Salvatore – Reagle (B) procedure

(a) Total CO emissions from mobile sources in the UK, 1980 – 1989

Explanatory Variable	*DW* Eq.(4) Original data	(A)	(B)	$2(1-r_1)$ Eq.(12) Original data	(A)	(B)
$x_t = 10^{-3}P_t$	1.235	1.446	0.430	1.305	1.795	0.193
$u_t = 10^3(P_t - P_1)$	0.655	1.594	1.777	1.409	1.988	1.875

(b) Total SO_2 emissions in the USA, 1980 - 1989

Explanatory Variable	*DW* Eq.(4) Original data	(A)	(B)	$2(1-r_1)$ Eq.(12) Original data	(A)	(B)
$v_t = Q_t - Q_1$	1.103	1.233	1.912	1.488	1.900	2.473
$z_t = Q_{10} - Q_t$	1.098	1.103	1.019	1.482	1.591	1.336

3. Further Considerations

3.1. Representation via Simple Linear Regression

The year – to – year upwards variation of substance levels (indicated by $b > 0$) in a small number of cases presumably signifies ineffective pollution control scenarios at this relatively early stage of ecological awareness. (The 1980's have been characterized, nonetheless, as the period of "waste minimization/reduction" [17]). Ineffectiveness, or lack of remedial action, can be inferred where minuscule annual variations lead to regressions with near zero R^2 values. Data collected in the USA during the 1980 – 1989 period are the most amenable to simple regression analysis, followed by the UK, West Germany, and The Netherlands. Data in Tables 8.6 – 8.20 [2], for Ireland have not been included in the analysis on account of their incompleteness except for SEI = 9, where the (full) annual observations yield $R^2 = 0.90$; $a = 0.2740$; $b = -0.030$; $DW = 1.545$, and indicate an approximately random residual distribution.

In principle, the quality of fit can be "improved" by means of polynomial (emission versus year) regressions, as in the case of total NO emissions from mobile sources in the Netherlands (SEI = 4M) with R^2 increasing from 0.483 in simple linear to 0.572 (quadratic), 0.723 (cubic), 0.823 (quartic) and 0.913 for a quintic fit. The practical desirability of a higher order polynomial fit is highly subjective, inasmuch as sensible improvement in prediction accuracy remains dubious. For instance, the amount of NO_2 in the air in the Netherlands in 1987: 0.354 megatonne by the quintic regression and 0.345 megatonne by the simple linear regression compared to the observed 0.350 megatonne is a case in point (due its low $R^2 = 0.483$, the simple linear regression has not been included in the data assembled in the tables).

3.2. Durbin – Watson Statistics for Observation Sizes $15 \leq n \leq 100$

As hinted at in Section 2.2, tighter rules for ascertaining the existence of autocorrelation via the Durbin – Watson statistic exist by establishing the position of a calculated DW value with respect to a lower (d_L) and an upper (d_U) critical value tabulated at a 5% and a 1% level of statistical significance [18 – 20]. The test is inconclusive for positive autocorrelation if $d_L \leq DW \leq d_U$, and for negative autocorrelation if $d_L \leq (4 - DW) \leq d_U$.

3.3. Representation via Autocorrelation – Free Regressions

Upon first attempt to remove autocorrelation, the regressions, rewritten as

$$\hat{Y}_t = -67.9321 + 0.2955\hat{Y}_{t-1} + 1.7900P_t - 0.5289P_{t-1} \tag{13}$$

for CO emissions in the UK (R^2 = 0.98; Table 5a; Procedure A; 2 (1 – r_1) = 1.988) and as

$$\hat{Y}_t = 220.0867 - 0.2513\hat{Y}_{t-1} - 0.3724P_t - 0.0936P_{t-1}; 2 \le t \le 10; P_1 = 114.23 \tag{14}$$

for SO_2 emissions in the USA (R^2 = 0.56; Table 5b; Procedure B; *DW* = 1.912), may be considered to be approximately free of autocorrelation. For the remainder in Table 5, repetitive applications of the two procedures, although feasible in principle [10], but not guaranteed to be successful, are not pursued here, except for a short illustration in the Appendix.

The finding of a fully successful "de - autocorrelating" scheme might well depend on the identification of properly chosen explanatory variables emerging from meteorological, climatic, demographic, etc. data sets. To the author's knowledge, such indicators were not (readily) available during the time period herein studied.

3.4. Verifying the Removal of Autocorrelation: The Von Neumann Ratio (VNR) Test

The statistic [5, 21, 22] defined as

$$VNR = \frac{\sum_{i=2}^{n}(y_i - y_{i-1})^2}{\sum_{i=1}^{n}(y_i - y_m)^2}; y_m \equiv \frac{1}{n}\sum_{i=1}^{n} y_i \tag{15}$$

when compared to tabulated values of the lower quantile of its null distribution, indicates positive autocorrelation if, for a general (y_1, y_2,…,y_n) observation set, *VNR* is *smaller* than the value of the *lower* quantile VNR_{cr} related to size *n*, at a specified level of statistical significance. For testing negative autocorrelation, *VNR* is replaced by *4 – VNR*. Applying Eq. (15) to the residuals associated with Eq.(13) and Eq.(14), *VNR* = 0.637 (UK) is significant at the 1% level (VNR_{cr} = 0.709). The *VNR* = 1.853 (USA) remains insignificant at any meaningful level of significance, due to its very large magnitude. In terms of the conventional theory of statistical hypothesis testing, rejection of the (null) hypothesis of residuals independence would carry an approximately 1% error in the former, and at least a 5% error ($VNR_{cr} = 1.062$) in the latter case.

Conclusion

In general, moving from descriptive to inferential statistics constitutes the first step in searching for quantitative relationships. The material presented in this chapter demonstrates the difficulties encountered when not all conceivably important variables and/or parameters are known, and numerical information is limited in extent. Nonetheless, appropriately (and often fortuitously) perceived conjectures can lead to at least partially sustainable inferences, until further comprehensive and consistent information in future pollution – control studies becomes available. The approach presented here, although by no means exhaustive, pinpoints the (potential) attractiveness of adopting a technique from another discipline for analyzing pollution data.

Acknowledgments

Facilities provided by the Natural Sciences and Engineering Research Council of Canada (NSERC), and the University of Waterloo are gratefully recognized.

Appendix

Illustration of a second attempt to remove autocorrelation in the time – series analysis of total CO emissions from mobile sources in the UK (1980 – 1989). Explanatory variable: $x_t = 10^{-3} P_t$

From Table 5a, the case of $DW = 1.446$; $2(1 - r_1) = 1.795$ is considered.

1. Cochrane – Ormutt procedure Since $r_1 = 1 - 1/1.795 = 0.1025$, second attempt variables $X_t^{''} = X_t^{'} - 0.1025X_t^{'}$ and $Y_t^{''} = Y_t^{'} - 0.1025Y_{t-1}^{'}$ are established. The resulting regression $\hat{Y}_t = 3.5324 - 0.02017x_t$ carries an R^2 value only slightly higher than zero, hence it would be pointless to pursue the matter further.

2. Salvatore – Reagle procedure Using the first attempt –based value set ($Y_t^{'}$; $X_t^{'}$) in obtaining regression parameters a_0 = - 50.0338; a_1 = 1.3961; a_2 = 0.5290; a_3 = 2.4x10^{-7} with R^2 = 0.976, the resulting regression $\hat{Y}_t^{'} = -35.7944 + 2.1440x_t^{'}$ carries R^2 = 0.845; DW = 1.554 and 2(1 – r_1) = 1.865, indicating a reasonable improvement with respect to the first attempt.

REFERENCES

[1] *The state of the environment in the European community* (92)23 Final, Vol. III, 27 March 1992, Brussels, Belgium.

[2] Kiely, G. *Environmental Engineering*. McGraw Hill: London, UK, 1997, pp. 334 – 3576.

[3] Durbin, J.; Watson, G. S. *Biometr., 1951, 38*, 159 – 178.

[4] Neter, J.; Wasserman, W.; Kutler, N. H. *Applied linear statistical models*. IRWIN: Homewood, IL, 1990, 3rd ed., pp. 491 – 494.

[5] Powell, F. C. *Statistical tables for the social, biological and physical sciences*. Cambridge Univ. Press: Cambridge, UK, 1982, p. 85.

[6] Bell, P.C.; Newson, E. F. P. *Statistics for business with spreadsheets: text and cases*. The Scientific Press: San Francisco, CA, 1992; 2nd ed. pp. 304 – 305.

[7] McClave J. T.; Benson, P. G.; Sincich, T. *Statistics for business and economics*. Pearson/Prentice Hall: Upper Saddle River, NJ, 2005, 9th ed., p. 1046.

[8] Fahidy, T. Z. *Chem. Eng. Educ., 2004, 38,* 22-25.

[9] Fahidy, T. Z. *Curr. Top. in Electrochem., 2014, 18,* 9 - 14.

[10] Neter, J.; Wasserman, W.; Kutler, N. H. loc. cit. pp. 496-500.

[11] Neter, J.; Wasserman, W.; Kutler, N. H. loc. cit. p. 197.

[12] Salvatore, D.; Reagle, D. (2002). *Statistics and econometrics.* Schaum's Outline Series, McGraw Hill: New York, NY, 2002, 2nd ed., pp. 208 – 209.

[13] Riviello, C.; Beretvas, S. N. *J. Mod. Stat. Meth. 2009, 8*, 469 – 477.

[14] Office for National Statistics UK:*Changes in UK population over the last fifty years* http://www.statisticsauthority.gov.uk;www.ons.gov.uk/ons/rel/pop-estimate/population – estimates.

[15] *US population by year-S & P 500 PE Ratio*, http://www.multpl./com/united-states-population/table.

[16] Neter, J.; Wasserman, W.; Kutler, N. H. loc. cit. p. 497.
[17] Curran, M. A. (2015). *Chem. Eng. Progr. 2015, 111*, 26 – 35.
[18] Neter, J.; Wasserman, W.; Kutler, N. H. loc. cit. pp. 140-1141.
[19] Powell, F. C., loc. cit. pp. 86-87.
[20] Keller, G.; Warrack, B. *Statistics for management and economics.* Duxbury: Pacific Grove, CA, 2000, pp. 304 -305.
[21] Hurt, B. I.; Newmann, J. V. *Ann. Mat. Stat. 1942, 13*, 207 – 214.
[22] Hurt, B. I. *Ann. Mat. Stat. 1942, 13*, 445 – 447.

BIOGRAPHICAL SKETCH

Name: Thomas Zoltan Fahidy
Distinguished Professor Emeritus; Professional Engineers of Ontario
Department of Chemical Engineering
University of Waterloo, 200 University Ave. W. Waterloo,
N2L 3G1, Ontario Canada

Following four years of undergraduate studies at the University of Chemical Technology Veszprém, Hungary (1952 – 1956) Dr. Fahidy obtained his B. Sc. (1959) and M. Sc. (1961) degrees from Queen's University (Canada), and his Ph. D. degree from the University of Illinois (Urbana – Champaign, 1965). Joining the University of Waterloo in late 1964 as an assistant professor, he was subsequently promoted to associate professor and professor in the Department of Chemical Engineering. From early 1965 to mid – 2003 he taught numerous undergraduate and graduate level courses, carried out supervised and personal research, and participated at various levels in academic administration until his formal/official retirement at the end of 2002.

He is the recipient of various professional honours. He was elected member of Sigma Xi, The Scientific Research Society (1965); Fellow of the Chemical Institute of Canada (1974); Fellow of the Royal Society of Canada (1997), Member of the General Assembly of the Hungarian Academy of Sciences (2002); Fellow of the American Institute of Chemical Engineers (2004), Fellow of the Electrochemical Society (2006), Fellow of the Canadian Academy of Engineering (2011).

Dr. Fahidy did path finding work in the area of deterministic and stochastic process dynamics and heat transfer theory applied to electrochemical processes, as well as in magneto - electrolysis (electrolysis in combined electric and magnetic fields). He was invited speaker at international conferences featuring magneto – hydrodynamics and related topics. His seminal paper entitled *Magnetoelectrolysis* in the Journal of Applied Electrochemistry Vol. 13, pp. 553 – 563: 1983 and his graduate- level text entitled *Principles of Electrochemical Reactor Analysis* (Elsevier, Amsterdam, The Netherlands, 1985) have received wide attention. His deep interest in ecological aspects of magnetism is illustrated by the invited chapter entitled *Magnetic Field Effects in Environmental Control Involving Electrolytes* (Studies in Environmental Science 59 Environmental Oriented Electrochemistry; ed. C. A. C. Sequeira, Elsevier, Amsterdam, 1994). He is author or coauthor of about 250 peer – refereed scientific/engineering journal articles and 25 book chapters. As a firm believer in the importance of cross – fertilization of scientific disciplines, Dr. Fahidy's interest shifted further, upon his "de iure" but not "de facto" retirement, towards the exploration of the utility of specific domains of applied probability theory and statistics in applied chemistry and chemical engineering, as exemplified by the following select publications: Basic Applications of the Analysis of Variance and Covariance in Electrochemical Science and Engineering (Chapter 2, Modern Aspects of Electrochemistry, eds. R. E. White, C. G. Vayenas, M. E. Gamboa – Aldeco, Springer Science Business Media, New York, NY, Vol.40, pp.37 – 74: 2007); Nonparametric Statistics in the Analysis of Certain Electrochemical Surface Phenomena: An Introduction (Chapter 6, New Developments in Electrodeposition and Pitting Research, ed. A. El Nemr, Transworld Research Network, Trivandrum, India, pp. 169 – 192: 2007); Applications of the Maximum Likelihood Technique to Certain Electrochemical Observations (Chapter 2, New developments in Electrodeposition and Pitting Research, ed. A. El Nemr, Transworld Research Network, Trivandrum, India, pp. 65 -81: 2007); Nonparametric Anova–Based Diagnostics of Electrochemical/ Electroanalytical Measurements: an Introduction (Chapter 6, Electroanalytical Chemistry Research Trends, ed. Kento Hayashi, Nova Science Publishers, New York, NY, pp. 143 – 161: 2009); An Application of Two – Sample Multivariate Statistical Techniques to the Analysis of Electrochemical Reactor Performance (Chapter 12, Applied Electrochemistry Chemistry Research and Applications Series, ed. V. G. Singh, Nova Science Publishers, New York, NY, pp. 437 – 452: 2010); Potential Applications of the Von Mises Distribution (VMD) of Circular Statistics to Chemical/Environmental Engineering (Chapter 9, Advances in Chemistry Research, ed. J. C. Taylor, Nova Science Publishers, New York, NY, Vol. 18, pp. 267 – 283: 2013); Potential Applications of Actuarial Pension Mathematics to Chemical Sciences

(Chapter 5, Advances in Chemistry Research, ed. J. C. Taylor, Nova Science Publishers, New York, NY, Vol. 26, pp. 77 -88: 2015).

His (temporary) external appointments included Shell Visiting Fellow at University College London (UK, 1968 – 1969); Canada – France Exchange Program Scientist at Université de Reims (1975 – 1976, 1977, 1983); Invited Professor, Ecole Polytechnique de Lausanne Switzerland (1986); Institut Universitaire de Technologie (IUT) St. Nazaire France (1990). He was the external examiner of three doctoral theses at the Université de Reims and the IUT St. Nazaire.

Dr. Fahidy's services rendered to professional organizations have been multifold. He was a member (1986 – 1988) and chairperson (1988 – 1989) of the Natural Sciences and Engineering Research Council of Canada (NSERC) Grant Selection Committee (chemical and metallurgical engineering) for the distribution of federal research grants at Canadian universities; member of a Chemical Institute of Canada (CIC) review team of the AECL Nuclear Waste Disposal Project (1995); vice – chair of a Canadian Engineering Accreditation Board (CEAB) Team at the University of Saskatchewan (1995); examiner for the Professional Engineers of Ontario (PEO); associate editor of the Canadian Journal of Chemical Engineering (1975 – 1981; 1990 – 1996) and recipient of its Certificate of Meritorious Service (1981); consultant with the Ontario Council of Graduate Studies (1988), and member of one of its appraisals committees (1989 – 1992). He was also Associate Dean of Graduate Studies (1992 – 1998), and member of the Senate (1991 – 1996) of the University of Waterloo. He is regularly featured in the Canadian Who is Who.

In: Air Pollution
Editor: Gerald L. Burns

ISBN: 978-1-63485-374-3

Chapter 2

SPATIAL ANALYSIS OF AIR POLLUTION FROM ROAD TRANSPORT WITHIN URBAN AREAS AND ITS RELATION TO HEALTH RISKS

Lubos Matejicek*

Institute for Environmental Studies, Charles University in Prague, Faculty of Natural Science, Prague, Czech Republic

ABSTRACT

This study extends the traditional research in air pollution and health risks by using spatial and temporal methods in geographical information system (GIS). The spatial interpolations are produced from point samples, which are represented by sites of monitoring stations, and by randomly generated sites with local estimates of urban air pollution by geographically weighted regression (GWR). The explanatory variables for estimates of urban air pollution, NO_2 and PM_{10}, are represented by the elevation, the distance to the nearest major road, and the ratio of the built-up area in the defined circles. The dependent variables contain annual average concentration of NO_2 and PM_{10} in Prague. The attached residuals and multiple R-squared indicate that each GWR model can partially

* Corresponding author: Lubos Matejicek. Institute for Environmental Studies, Charles University in Prague, Faculty of Natural Science, Benatska 2, 128 01, Prague, Czech Republic. E-mail: lmatejic@natur.cuni.cz.

explain the variation of concentrations in relation to the selected exploratory variables. Studying the relations between urban air pollution, such as NO_2 and PM_{10}, and lung cancer incidence are provided separately for ten districts of the city, independently for males and females. In spite of that the spatial analysis of urban air pollution on a reduced number of monitoring stations and limited available spatial datasets can be beneficial for decision-making processes, relations of the lung cancer incidence to urban air pollution indicate weak or even non-existed associations. The available datasets about the lung cancer incidence are mostly based on aggregated information for each district in the city. Thus, there is a little evidence about original sites of the lung cancer incidence, which documents many studies that are focused on assessing the public health impacts by urban air pollution. Presented spatial and temporal analysis of air pollution from road-transport within the urban areas represents a new way of research that is based on integration of urban air pollution monitoring, understanding potential human exposure, and advanced modeling tools implemented in ArcGIS. It can be used for risk assessment, decision-making processes and urban planning. It is anticipated that the use of advanced methods of exploratory spatial data analysis will bring new progress into the studied processes and new insights for urban planning and health protection.

Keywords: urban air quality, road transport, geographically weighted regression, spatial and temporal modeling, lung cancer

1. INTRODUCTION

Urban air pollution caused by increased vehicle usage is one of the major problems currently confronting the world's population. Thus, exploring air quality information in relation to mobile sources is important in understanding potential human exposure and impacts in health. Increasing evidence from epidemiological studies indicates association of exposure to ambient air pollution with adverse health effects [1, 2]. Many studies are focused on exceedances and variation of the short- and long-term objectives for PM_{10} and exceedance of the long-term objectives for NO_2 [3, 4, 5, 6]. Adverse effects of urban air pollution include pre-mature mortality and morbidity from cardiovascular and respiratory causes with relation to lung cancer incidence. It is mostly assessed using statistical models, such as regression and correlation, which are extended by dispersion modelling techniques and traffic indicator variables [7, 8]. A number of studies deal with development of modelling tools for mapping urban traffic emissions and air quality. In additions to the EPAs models, other models are developed to bring together information on emissions, atmospheric transport between sources and exposed areas and emission control measures [9, 10]. Some tools for air quality data analysis are developed as extensions of widespread statistical programs in order to complement standard functions and methods. For example open air, which is an R package for air quality data analysis, can manage and evaluate air pollution data together with a wide range tools accessible in other R packages [11]. For more complex analysis of spatio-temporal phenomena, geographic information systems (GIS) are used in order to integrate spatial map layers together with information on emissions and exposure assessment [12, 13].

To protect the quality of life and to minimize potential human exposure of European citizens, the European Commission established limit values for NO_2, PM_{10} and other air pollutants in the Air Quality Daughter Directive 1999/30/EC (EC, 1996). It was replaced in 2008 by the new Directive 2008/50/EC (EC, 2008). Because existing air quality guidelines for NO_2 and PM_{10} are exceeded at many locations throughout Europe, there is increased interest at the EU policy level in the exposure effects and their impacts in health. Thus, many studies are focused on implementation of cost-effective control of air quality [14], classifications of road environment for assessment of human exposure [15], and integration of new approaches in the framework of European policy [16]. In the framework of the ESCAPE study (Study of Cohorts for Air Pollution Effects), the measurements of NO_2, NO_x and PM_{10} concentrations in 36 study areas during one year period indicated variation of air pollution between and within 36 European study areas in 15 different European countries [3]. The further assessment was also focused on the difference of measured concentrations at traffic stations versus urban background stations as a source of within study area spatial variation [17, 18].

2. METHODS

A number of studies have been presented for a few decades to advance spatio-temporal research in analysis of urban air pollution. Generally, Oxley identified three key challenges for issue-driven integrated assessment models, which deal with (i) capturing the dynamics of behavioural responses to value-driven policy mechanisms designed to promote non-technical abatement measures; (ii) the development of new tools to handle the delay times, inter-dependencies and multi-pollutant effects of abatement measures; and (iii) the identification of boundaries between knowledge domains and the effective transfer of multi-disciplinary scientific and socio-cultural knowledge to an increasingly trans-disciplinary policy domain [19]. More specific studies are focused on implementation of spatial interpolation methods into air pollution research [20]. Other contributions deal with using of case oriented tools, such as kriging techniques or geographically weighted regression (GWR), for processing of urban air pollution information in GIS [21, 22].

This paper is focused on spatio-temporal analysis of air pollution from urban road-transport and its relation to human health using GWR and interpolation techniques. GWR is a fairly recent contribution to modelling spatially heterogeneous processes [23]. The underlying idea of GWR is that parameters can be estimated anywhere in the area of interest given a dependent variable and a set of explanatory variables which have been measured at sites whose locations are known. GWR provides a local model for the dependent variable by fitting a regression equation to every feature in the dataset. It constructs these separate equations by incorporating the dependent and explanatory variables of features falling within the bandwidth of each target feature. The shape and size of the bandwidth is dependent on setting the kernel type, bandwidth method, distance, and number of neighbours. The employed spatial GWR and interpolation algorithms are implemented in ArcGIS together with exploratory spatial data analysis [24], and can be used for prediction of air pollution in urban areas [25].

3. Input Datasets for Analysis of Air Pollution and Potential Human Exposure

The area of interest is represented by the territory of Prague, the capital and the largest city of the Czech Republic (area 496 km^2, population 1.25 million). The exceeding of the limit values relates with significant traffic loads for the whole year and seasonally with heating households. Each year, the limit value for the average 24-hour limit of suspended particles PM_{10} and the annual limit for nitrogen dioxide NO_2 are exceeded on a few stations [26]. The map schema of the target territory complemented by map layers with built-up sites and major roads is in Figure 1. The map schema containing a digital elevation model (DEM) and other layers such as major roads, stations for monitoring of urban air pollution and sites for prediction of urban air pollution by GWR is in Figure 2.

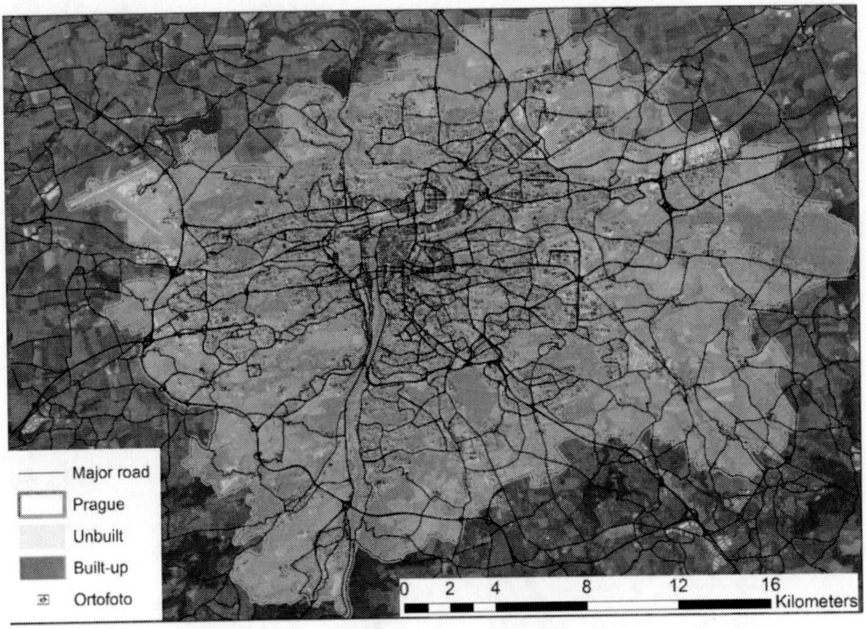

Figure 1. The target territory with major roads, built-up sites, and ortofoto in the background.

The configuration of a monitoring network is represented by17 stations that measure 12 main air pollutants. Datasets are recorded in one hour periods, and archived in the database of the Air Quality Information System of the Czech Republic. In this paper, mapping of the urban air pollution is based on two pollutants (concentration of NO_2 and PM_{10}), which were measured in 2011-2013. The configuration of 100 additional (randomly generated) monitoring sites for prediction of urban air pollution (concentrations of NO_2 and PM_{10}) by GWR is included in a random point layer.

4. RESULTS

Finally, a list of the GWR explanatory variables for estimates of urban air pollution (concentration of NO_2 and PM_{10}) included: the elevation, the distance to the nearest major road and the ratio of built-up area in the defined circle. The diameter of a circle for the ratio of built-up area was 300 meters for NO_2, and 150 meters for less mobile particles PM_{10}. The values of selected explanatory variables are together with the observed values of dependent variables (concentration of PM_{10} and NO_2 for each year in 2011-2013) in Table 1. The observed values and the corresponding predicted values for dependent variables estimated by GWR (concentrations of PM_{10} and NO_2 for each year in 2011-2013) are in Table 2. Each observed value and its predicted value by GWR is complemented by the residual value, in order to evaluate an unexplained portion of the dependent variable, which is also one measure of the model fit. The multiple *R*-squared for each year indicates that the GWR model can only partially explain the variation in the dependent variable. In case of urban air pollution by NO_2, the multiple *R*-squared is $R^2_{2011} = 0.64$, $R^2_{2012} = 0.57$ and $R^2_{2013} = 0.62$ for 2011, 2012 and 2013, respectively. In case of urban air pollution by PM_{10}, the multiple *R*-squared is $R^2_{2011} = 0.35$, $R^2_{2012} = 0.42$ and $R^2_{2013} = 0.63$ for 2011, 2012 and 2013, respectively.

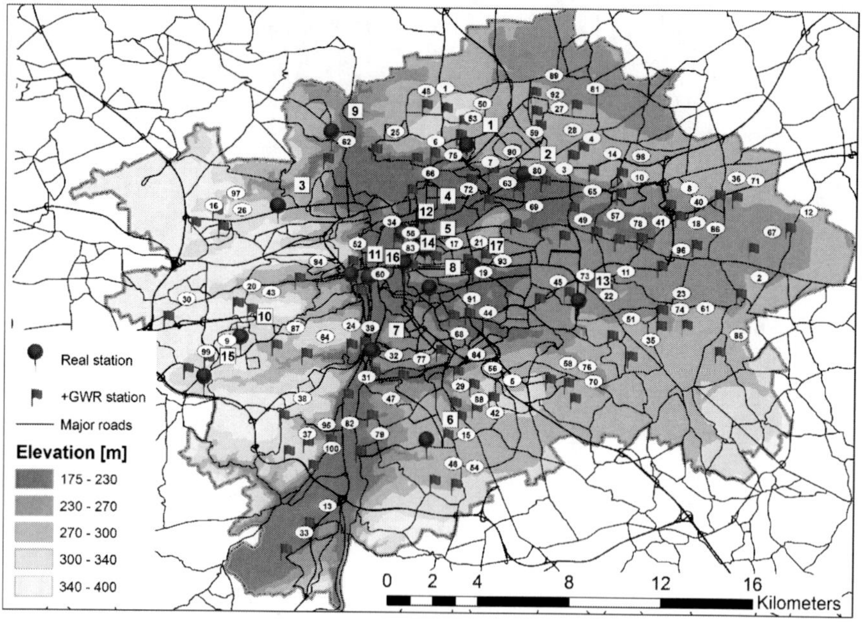

Figure 2. The target territory complemented by a digital elevation model, major roads, stations for monitoring of air pollution (Real station: 1-17) and sites for prediction of air pollution by GWR (+GWR station: 1-100).

Table 1. Values of explanatory variables (elevation, distance to the nearest major road and ratio of the built-up area) and the values of dependent variables (concentration of PM_{10} and NO_2). A few values of were substituted by values from 2011 to complete a dataset (marked by grey background) [27]

Station	Elevation	Distance	Built-up	Built-up	Annual average value			Annual average value		
ID	[m]	[m]	300 m	100 m	PM_{10} [μg·m^{-3}]			NO_2 [μg·m^{-3}]		
					2011	2012	2013	2011	2012	2013
1	275	288	0.10	0.01	22.2	20.1	19.9	25.8	23.7	23.4
2	209	10	0.27	0.15	31.1	27.2	24.1	40.8	39.3	38.6
3	314	54	0.12	0.09	27.3	38.1	31.0	29.1	31.3	28.6
4	200	8	0.37	0.54	32.0	32.5	32.4	35.9	28.0	32.0
5	250	125	0.22	0.06	26.9	23.8	22.7	30.6	28.3	25.9
6	300	56	0.02	0.00	27.7	27.9	29.4	21.1	21.3	21.7
7	210	52	0.08	0.09	23.3	19.6	19.2	30.0	30.8	31.1
8	200	5	0.26	0.12	32.4	26.1	30.3	32.4	32.4	32.4
9	261	212	0.07	0.10	27.5	25.2	26.5	22.0	23.2	20.3
10	325	474	0.13	0.10	26.6	24.6	25.2	25.7	24.4	25.7
11	207	0	0.28	0.16	39.0	30.3	27.6	47.6	42.7	39.8
12	200	527	0.67	0.59	30.5	28.9	27.4	37.9	36.7	34.3
13	274	40	0.27	0.17	31.2	28.9	27.4	36.6	36.6	34.3
14	230	4	0.58	0.49	29.6	28.2	29.8	69.3	57.8	53.3
15	324	141	0.17	0.22	35.0	29.4	30.9	32.5	30.6	31.7
16	200	0	0.53	0.71	35.8	29.8	34.2	40.4	38.1	35.9
17	241	155	0.21	0.17	23.2	19.9	21.4	44.6	44.6	40.4

The GWR model for each pollutant and each year is used to create new predictions. In order to create a more complex dataset for spatial interpolations, a random point dataset was generated in the inner area that is one kilometre distant from the territorial boundaries of Prague. After the creation of the random point dataset, its values of concentration for urban air pollution (NO_2 and PM_{10}) for each year in 2011-2013 were predicted by GWR as dependent variables in relation to explanatory variables based on the same local urban characteristics such as the elevation, the distance to the nearest major road, and the ratio of built-up area in the defined circle. Spatial interpolations based on the observed values at 17 stations, and an extension of other 100 randomly generated sites with the predicted values by GWR for PM_{10} and NO_2 are in Figure 3 and Figure 4, respectively.

Table 2. The observed values of independent variables and the corresponding predicted values of independent variables by GWR (concentration of PM_{10} and NO_2 for each year in 2011-2013) complemented by the residuals and the multiple *R*-squared for each pollutant and each year in 2011-2013

	NO_2 [μg·m^{-3}]									PM_{10} [μg·m^{-3}]								
ID	2011 ($R^2_{2011} = 0.64$)			2012 ($R^2_{2012} = 0.57$)			2013 ($R^2_{2013} = 0.62$)			2011 ($R^2_{2011} = 0.35$)			2012 ($R^2_{2012} = 0.42$)			2013 ($R^2_{2013} = 0.63$)		
	Obs.	Pr.	Res.	Obs.	Pr.	Res.	Obs.	Pr.	Res.	Obs.	Pr.	Res.	Obs.	Pr.	Res.	Obs.	Pr.	Res.
1	25.8	24.0	1.8	23.7	24.7	-1.0	23.4	24.3	-0.9	22.2	25.7	-3.5	20.1	23.5	-3.4	19.9	22.7	-2.8
2	40.8	37.9	2.9	39.3	35.6	3.7	38.6	34.4	4.2	31.1	29.7	1.4	27.2	25.3	1.9	24.1	25.2	-1.1
3	29.1	33.0	-3.9	31.3	31.6	-0.3	28.6	30.4	-1.8	27.3	29.4	-2.1	38.1	30.0	8.1	31.0	29.3	1.7
4	35.9	42.7	-6.8	28.0	39.2	-11.2	32.0	37.6	-5.6	32.0	33.8	-1.8	32.5	30.4	2.1	32.4	31.5	0.9
5	30.6	33.6	-3.0	28.3	32.1	-3.8	25.9	31.1	-5.2	26.9	27.8	-0.9	23.8	24.9	-1.1	22.7	24.3	-1.6
6	21.1	27.2	-6.1	21.3	27.3	-6.0	21.7	26.7	-5.0	27.7	28.2	-0.5	27.9	27.8	0.1	29.4	26.9	2.5
7	30.0	27.5	2.5	30.8	27.8	3.0	31.1	27.5	3.6	23.3	28.5	-5.2	19.6	23.9	-4.3	19.2	23.6	-4.4
8	32.4	37.5	-5.1	32.4	35.3	-2.9	32.4	34.2	-1.8	32.4	29.2	3.2	26.1	24.3	1.8	30.3	24.1	6.2
9	22.0	24.4	-2.4	23.2	25.1	-1.9	20.3	24.8	-4.5	27.5	27.4	0.1	25.2	25.0	0.2	26.5	24.5	2.0
10	25.7	21.9	3.8	24.4	22.9	1.5	25.7	22.2	3.5	26.6	25.1	1.5	24.6	25.4	-0.8	25.2	24.6	0.6
11	47.6	38.9	8.7	42.7	36.3	6.4	39.8	35.1	4.7	39.0	29.8	9.2	30.3	25.4	4.9	27.6	25.3	2.3
12	37.9	43.1	-5.2	36.7	39.1	-2.4	34.3	36.9	-2.6	30.5	28.9	1.6	28.9	24.6	4.3	27.4	25.7	1.7
13	36.6	39.7	-3.1	36.6	36.7	-0.1	34.3	35.1	-0.8	31.2	30.0	1.2	28.9	28.9	0.0	27.4	28.6	-1.2
14	69.3	54.6	14.7	57.8	48.0	9.8	53.3	45.3	8.0	29.6	33.5	-3.9	28.2	31.5	-3.3	29.8	32.3	-2.5
15	32.5	33.3	-0.8	30.6	31.7	-1.1	31.7	30.4	1.3	35.0	29.9	5.1	29.4	31.2	-1.8	30.9	30.8	0.1
16	40.4	51.0	-10.6	38.1	45.4	-7.3	35.9	43.1	-7.2	35.8	35.7	0.1	29.8	32.9	-3.1	34.2	34.5	-0.3
17	44.6	32.0	12.6	44.6	31.0	13.6	40.4	30.0	10.4	23.2	28.6	-5.4	19.9	25.6	-5.7	21.4	25.4	-4.0

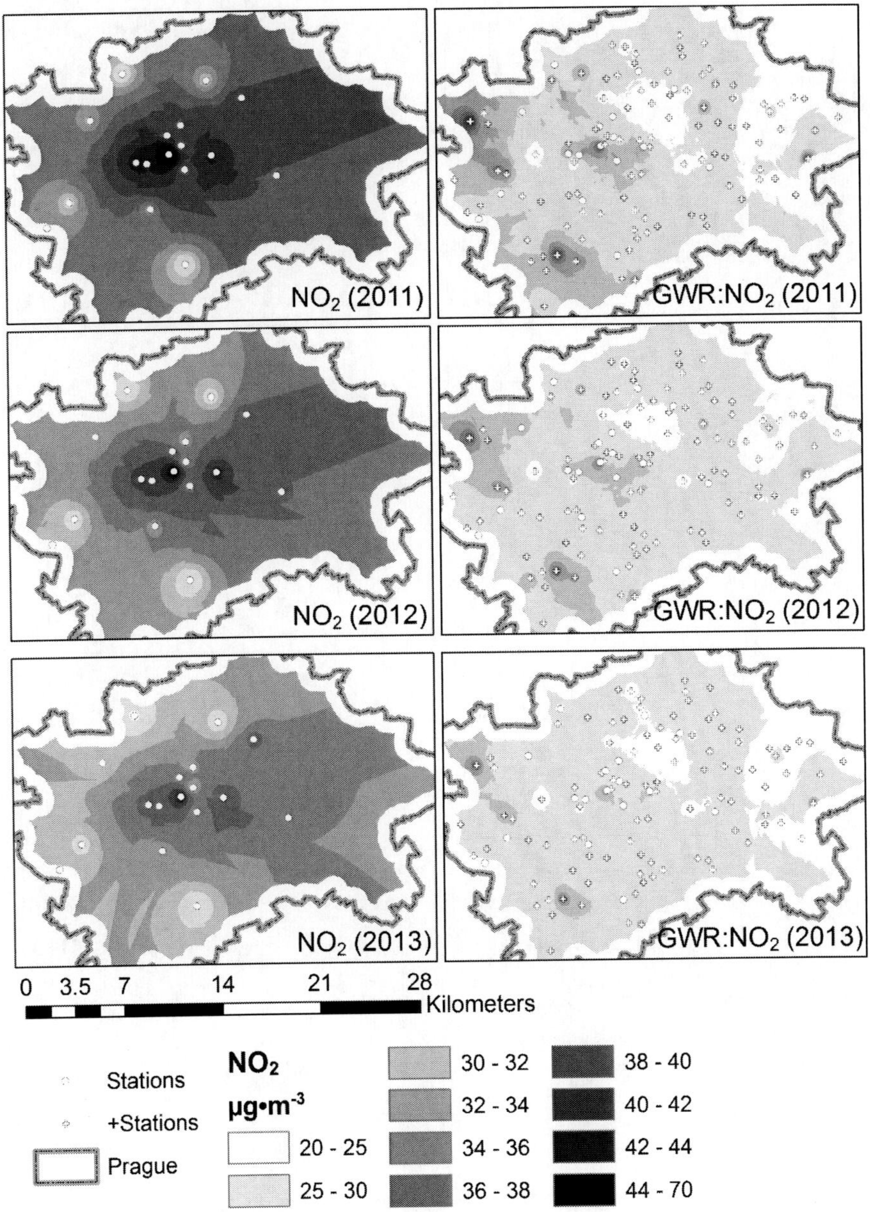

Figure 3. Spatial interpolations based on the observed values at 17 stations (on the left side), and an extension of other 100 randomly generated sites with the predicted values by GWR (on the right side) for NO_2. The spatial interpolations are created with IDW, independently for each year in 2011-2013.

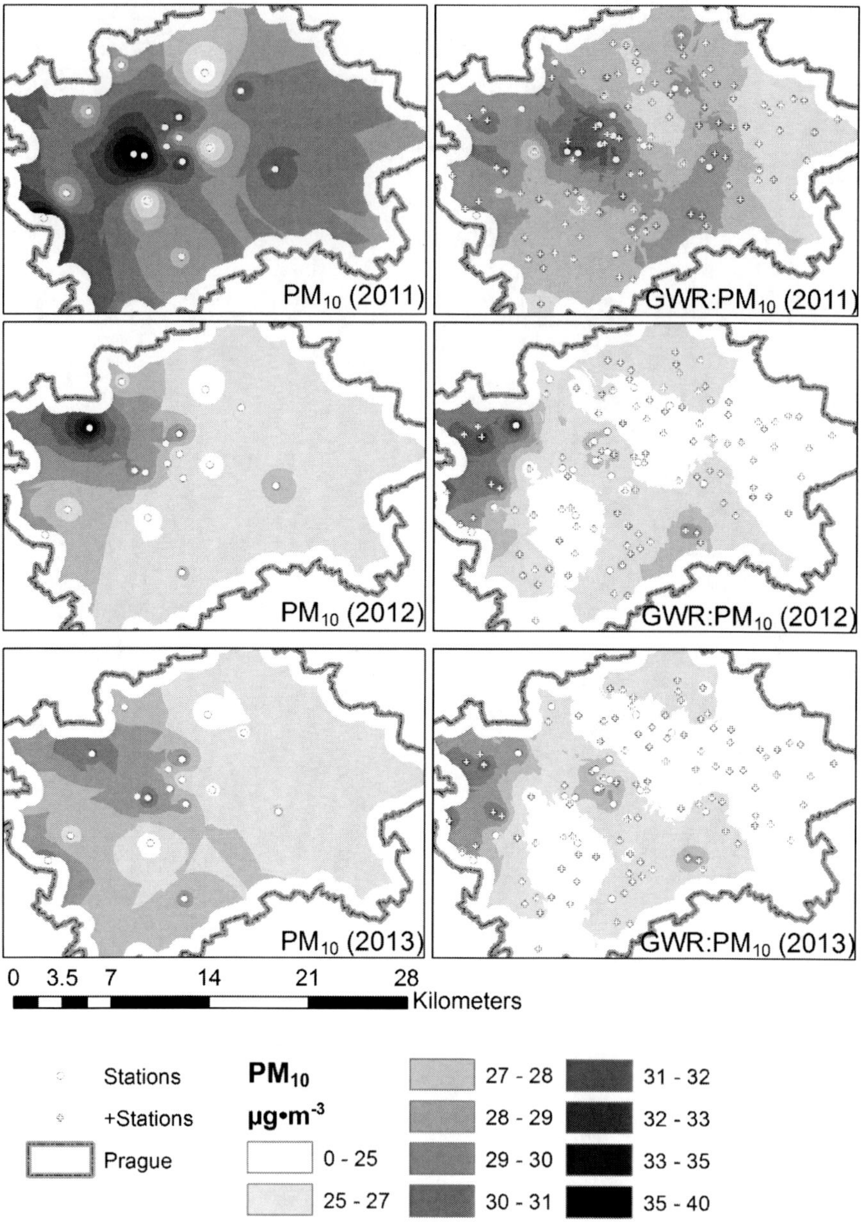

Figure 4. Spatial interpolations based on the observed values at 17 stations (on the left side), and an extension of other 100 randomly generated sites with the predicted values by GWR (on the right side) for PM10. The spatial interpolations are created with IDW, independently for each year in 2011-2013.

The spatial interpolations of concentration are created independently for each year in 2011-2013 for both the pollutants, PM_{10} and NO_2. They contain continuous surfaces, predicting the value of PM_{10} and NO_2 concentrations for every location in the inner territory of Prague. Former interpolation maps based on 17 point measurements can bring only a basic view about spatial distribution of pollutants. The exploration of urban air pollution in a more distant sites based on these spatial interpolations can be confusing as is illustrated in Figure 3 and Figure 4 in comparison with an extension of other 100 sites with the predicted values by GWR. The actual interpolation maps allow to incorporate more of the local spatial relationships that are discovered among the points and exploratory variables. In order to compare interpolation maps based on extended spatial datasets by GWR, the basic interpolation technique (Inverse Distance Weighted - IDW) is used, because the original datasets based only on the values from existing stations did not meet all the requirements for considered geostatistical methods, such as ordinary, simple or indicator kriging.

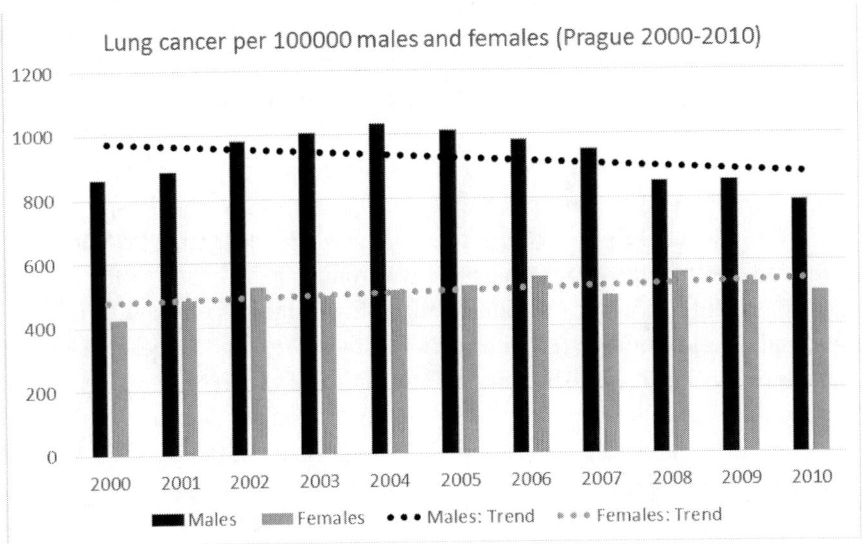

Figure 5. The sums of the notified cases of the lung cancer in all districts per 100000 persons, independently for males and females and for each year in 2000-2010 [28].

The next part of this study deals with relations of urban air pollution from road-transport to the lung cancer incidence in 2000-2010. Available datasets contain notified cases of malignant neoplasms by territorial districts independently for males and females for each year in 2000-2010. The sums of notified cases of the lung cancer in all districts per 100000 persons, independently for males and females and for each year in 2000-2010 are illustrated in Figure 5. Both the time series are complemented by trend lines that indicate the slightly decreasing lung cancer incidence for males in a last few years, and the slightly increasing lung cancer incidence for females. It can indicate a more significant relation to increased smoking of females in the last decades instead of a relation to the urban air pollution such as for example air pollution from road-transport within the urban areas. Thus, the exploration of potential associations with urban air pollution, especially concentrations of NO_2 and PM_{10}, is explored by GIS and discussed in the next part.

5. Discussion

The results in Table 2 show the residuals between observed and estimated concentration of pollutants for each station and each year in 2011-2013. The attached multiple *R*-squared for each year indicates that the GWR models can only partially explain the variation in the dependent variables (concentration of NO_2 and PM_{10}) in relation to the elevation, the distance to the nearest major road, and the ratio of built-up area in a defined circle. The multiple *R*-squared for NO_2 is: $R^2_{2011} = 0.64$, $R^2_{2012} = 0.57$ and $R^2_{2013} = 0.62$ for 2011, 2012 and 2013, respectively. The multiple *R*-squared for PM_{10} is lower: $R^2_{2011} = 0.35$, $R^2_{2012} = 0.42$ and $R^2_{2013} = 0.63$ for 2011, 2012 and 2013, respectively. Regression coefficients show a negative relation to the distance to the nearest major road, and a positive relation to the ratio of built-up area in a defined circle, and the elevation. A few sites indicate higher residual values, which can be caused by local disturbances of exploratory variables, and by other unknown local effects. In order to improve model predictions, a higher number of monitoring stations and more complex environmental datasets are needed for more accurate predictions and exploratory sensitivity analysis that can clarify other local effects such as transport intensity and local urban meteorological variables: the average wind speed, the predominant wind direction and the surface temperature.

In terms of the GWR processing, the methods implemented in ArcGIS 10.x were used together with other spatial data management tools for creation of the GIS project and for pre-processing of data sources from monitoring stations and thematic map layers for estimates of the elevation, the distance to the nearest major road, and the ratio of built-up area in a defined circle. Due to a low number of monitoring stations for model calibrations and testing of various optimization scenarios, the GWR most important parameters were set to the fixed kernel type, and to the corrected Akaike Information Criterion (AICc) for optimization of the bandwidth method parameters. The testing of various GWR parameters was provided by Robinson [29] for increasing the accuracy of NO_2 pollution mapping using GWR in relation to dispersion modelling outputs. Spatial residual analysis for GWR and a few other modelling techniques, such as linear mixed model (LMM), generalized additive model (GAM), multi-layer perceptron (MLP) and neural network, is described by Zhang [30]. In comparison to other methods, the GWR models had more desirable spatial distributions of residuals, and coefficients at each location in the study area produced more accurate predictions for the dependent variable. Investigating relationships between urban surface temperature and environmental factors using GWR was described by Li [31]. The exploratory variables included the distance between city and highway, patch richness density of forestland, wetland, built-up land and unused land and topographic factors such as elevation and slope. For reference, the ordinary least squares (OLS) model, a global regression technique, was also employed, using the same dependent and explanatory variables as in the GWR model. The results indicated that the GWR model provided a better fit and local detailed information about the spatial variation of dependent variable. The spatial non-stationary and scale-dependent factors were also discussed, which could be partly applied to the exploration of air pollution in the urban environment. In this paper, the spatial scale is limited by the inner area that is one kilometre distant from the territorial boundaries of Prague in order to provide local estimates for the exploratory variables from available urban spatial datasets (the circles with a diameter 300 meters for NO_2, and the circles with a diameter 150 meters for less mobile particles PM_{10} in the final GIS project). The spatial resolution for a city area (496 km^2) is 30 meters, which is sufficient for the output accuracy and the calculation efficiency. In addition to the selected exploratory variables (elevation, distance to the nearest major road and ratio of built-up area in a defined circle), other driving factor were included into the original set of explanatory variables. Likewise in the previous study about urban the surface temperature and the environmental factors [31], there was an idea to use satellite images for classification of the urban surface, and for estimates of the surface temperature from thermal bands. An applied spatial research for asthma hospitalizations was carried by Ayres-Sampaio [32]. The study found out that the surface temperature, normalized differential vegetation index (NDVI) and concentration of NO_2

had consistent relationships with asthma in all seasons. Other factors, such as relative humidity interpolated from meteorological stations and PM_{10} indicated very weak or non-existent associations. In this paper, the utilization of the NDVI or the surface temperature from satellite images caused local multicollinearity problems in the GWR models in spite of an effort to improve optimization of the GWR models. The reasons can be caused by a reduced number of monitoring stations that are often near the main crossroads, which give a limited possibility to explore the relationship between the dependent variables and the mentioned exploratory variables, such as the surface temperature, the NDVI or the humidity. But, exploring air quality information in relation to the potential human exposure and the impacts in health through the environmental factors captured by satellite images will be very beneficial in the future due to the improving technology of satellite systems and the increased availability of images [33].

In spite of that the spatial analysis of urban air pollution on a reduced number of monitoring stations and limited available spatial datasets can be beneficial for decision-making processes, relations of the lung cancer incidence to urban air pollution indicate weak or even non-existed associations. The datasets contain only information about notified cases of malignant neoplasms by territorial districts independently for males and females in 2000-2010. In order to provide spatial exploratory analysis, the time series of notified cases of lung cancer per 100000 males and females in ten districts complemented by the spatial interpolation of NO_2 (in 2011) and the spatial interpolation of PM_{10} (in 2011) is illustrated in Figure 6 and Figure 7, respectively. The available datasets about the lung cancer incidence are based on aggregated information for each district in the city. Thus, there is no evidence about original sites of the lung cancer incidence. In order to explore sites of the potential human exposure and the impacts in health, urban air pollution for NO_2 in 2011 and PM_{10} in 2011 for built-up sites is illustrated in Figure 8 and Figure 9, respectively. The basic map layer with built-up sites shows the concentration for each pollutant. In addition to a projection of the concentration (NO_2 in 2011 and PM_{10} in 2011), there is a map schema summarising the concentration only in the built-up sites for each district and a map schema of the lung cancer incidence. Because the actual available datasets for the lung cancer incidence are available for each year to 2010 and the most compact information about air pollution is for each year in 2011-2013, input data for the lung cancer incidence are from 2010 and input data for spatial interpolations of urban air pollution are from 2011. In this case, the spatial exploratory data analysis based on a combination of the listed output map layers was preferred to using standard statistics or multivariate spatial statistical tools, which can give confusing result due to the weak or non-existing associations. Thus, in spite of that adverse effects of urban air pollution have been explored for a few decades [34, 35], the direct relations described by deterministic models can be hardly accomplished. It documents many studies that are focused on assessing the public health impacts of urban air pollution [36, 37, 38].

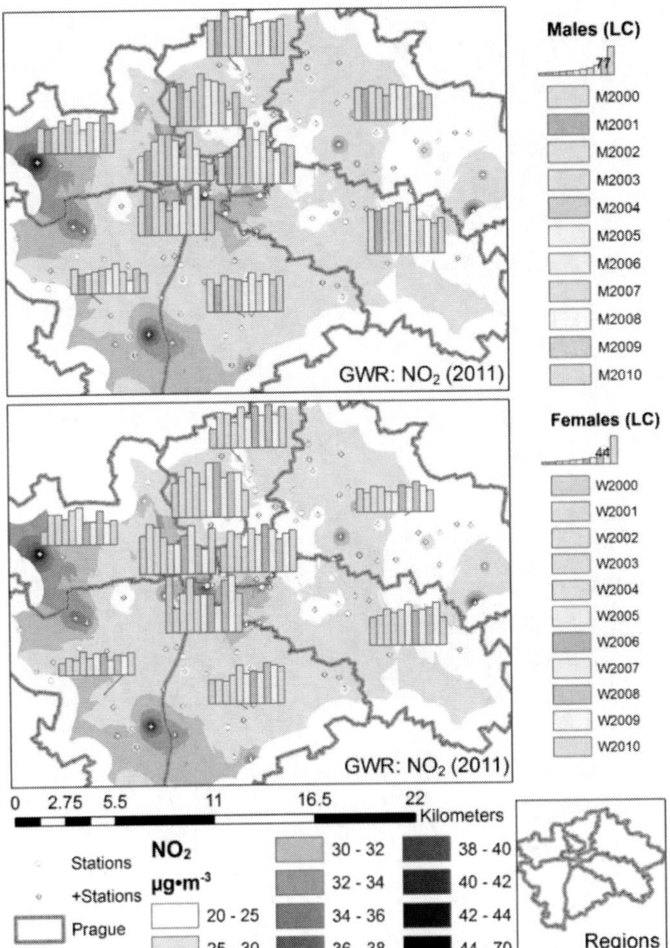

Figure 6. The time series of notified cases of lung cancer per 100000 males (in the upper part) and females (in the bottom part) for each year in 2000-2010 in ten districts complemented by spatial interpolation of NO_2 for 2011.

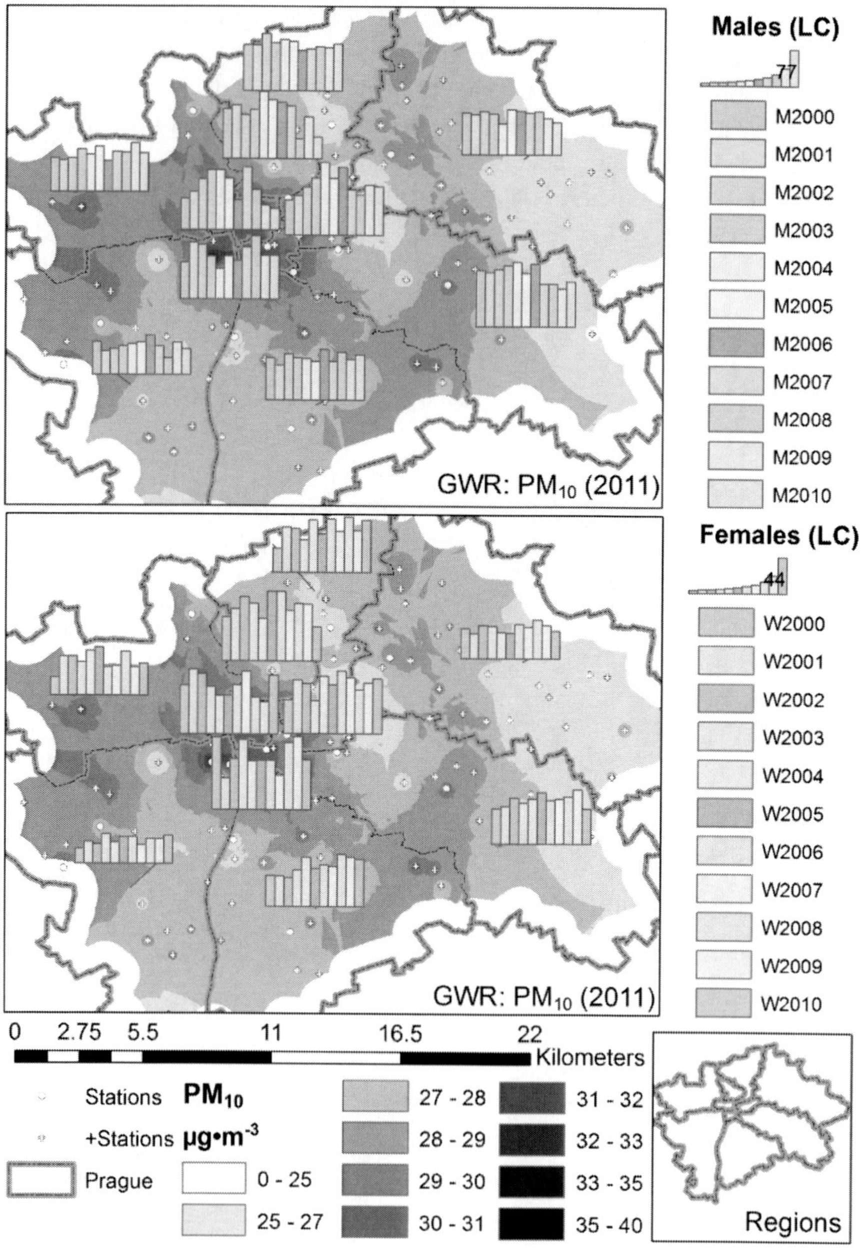

Figure 7. The time series of notified cases of lung cancer per 100000 males (in the upper part) and females (in the bottom part) for each year in 2000-2010 in ten districts complemented by spatial interpolation of PM_{10} for 2011.

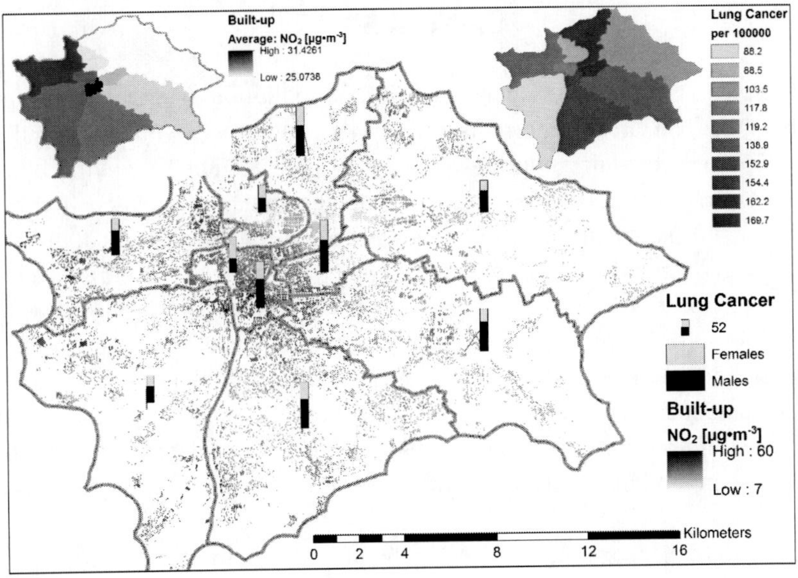

Figure 8. The map schema of air pollution for NO_2 in 2011 at built-up areas complemented by a map schema summarising by the mean the concentration only in the built-up sites for each district, and a map schema with lung cancer incidence.

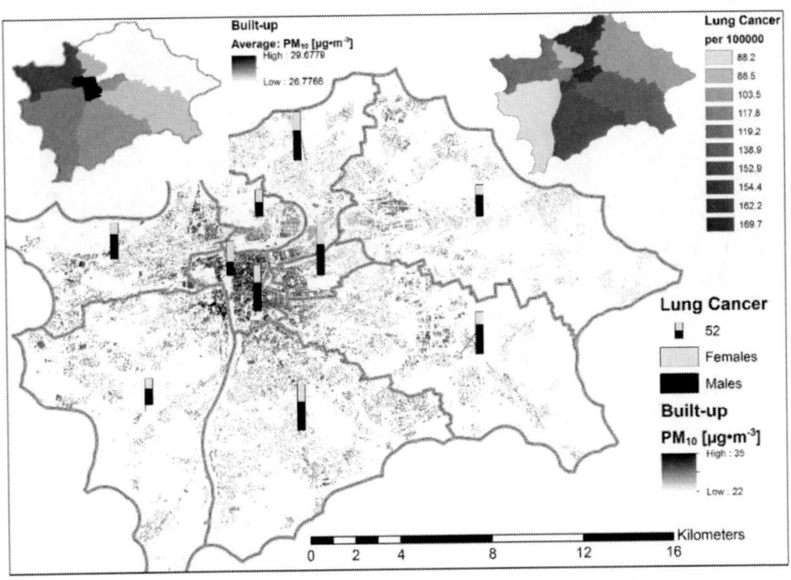

Figure 9. The map schema of air pollution for PM_{10} in 2011 at built-up areas complemented by a map schema summarising by the mean the concentration only in the built-up sites for each district, and a map schema with lung cancer incidence.

Conclusion

Presented spatio-temporal analysis of air pollution from road-transport within the urban areas represents a new way of research that is based on integration of urban air pollution monitoring, understanding potential human exposure and advanced modelling tools implemented in ArcGIS. It can be used for risk assessment, decision-making processes and urban planning. It is based on GWR methods that are complemented by other spatial analysis in GIS environment. The study is accommodated to limited input datasets, which can be useful for implementation in many similar urban areas in the world. On the other side, the lack of detail information about urban air pollution and lung cancer incidence enables to explore only basic associations and partial relations to lung cancer incidence. However, further developments consisting in spatio-temporal modelling of more complex datasets are planned in the future.

Considering the spatial scale and time period of a few years, an extended range of measurements and observations will be required to obtain more complex datasets for a new advanced spatio-temporal models. It is anticipated that the use of advanced methods of exploratory spatial data analysis will bring new progress into the studied processes and new insights for urban planning and health protection.

Acknowledgments

The geostatistical research was processed in the GIS Laboratory at the Faculty of Science, Charles University in Prague and was supported in the framework of FRVS project 131/2014/A/a.

References

[1] Beelen, R., Raaschou-Nielsen, O., Stafoggia, M., Andersen, Z. J., Weinmayr, G., Hoffmann, B., Wolf, K., Samoli, E., Fischer, P., Nieuwenhuijsen, M., Vineis, P., Xun, W. W., Katsouyanni, K., Dimakopoulou, K., Oudin, A., Forsberg, B., Modig, L., Havulinna, A. S., Lanki, T., Turunen, A., Oftedal, B., Nystad, W., Nafstad, P., De Faire, U., Pedersen, N. L., Östenson, C.-G., Fratiglioni, L., Penell, J.,

Korek, M., Pershagen, G., Eriksen, K. T., Overvad, K., Ellermann, T., Eeftens, M., Peeters, P. H., Meliefste, K., Wang, M., Bueno-de-Mesquita, B., Sugiri, D., Krämer, U., Heinrich, J., de Hoogh, K., Key, T., Peters, A., Hampel, R., Concin, H., Nagel, G., Ineichen, A., Schaffner, E., Probst-Hensch, N., Künzli, N., Schindler, C., Schikowski, T., Adam, M., Phuleria, H., Vilier, A., Clavel-Chapelon, F., Declercq, C., Grioni, S., Krogh, V., Tsai, M.-Y., Ricceri, F., Sacerdote, C., Galassi, C., Migliore, E., Ranzi, A., Cesaroni, G., Badaloni, C., Forastiere, F., Tamayo, I., Amiano, P., Dorronsoro, M., Katsoulis, M., Trichopoulou, A., Brunekreef, B., Hoek, G., 2014. Effects of long-term exposure to air pollution on natural-cause mortality: an analysis of 22 European cohorts within the multicentre ESCAPE project. *Lancet* 383, 785–95.

[2] Jephcote, C., Chen, H., 2013. Geospatial analysis of naturally occurring boundaries in road-transport emissions and children's respiratory health across a demographically diverse cityscape. *Soc. Sci. Med.* 82, 87–99.

[3] Cyrys, J., Eeftens, M., Heinrich, J., Ampe, C., Armengaud, A., Beelen, R., Bellander, T., Beregszaszi, T., Birk, M., Cesaroni, G., Cirach, M., de Hoogh, K., De Nazelle, A., de Vocht, F., Declercq, C., Dėdelė, A., Dimakopoulou, K., Eriksen, K., Galassi, C., Grąulevičienė, R., Grivas, G., Gruzieva, O., Gustafsson, A. H., Hoffmann, B., Iakovides, M., Ineichen, A., Krämer, U., Lanki, T., Lozano, P., Madsen, C., Meliefste, K., Modig, L., Mölter, A., Mosler, G., Nieuwenhuijsen, M., Nonnemacher, M., Oldenwening, M., Peters, A., Pontet, S., Probst-Hensch, N., Quass, U., Raaschou-Nielsen, O., Ranzi, A., Sugiri, D., Stephanou, E. G., Taimisto, P., Tsai, M.-Y., Vaskövi, É., Villani, S., Wang, M., Brunekreef, B., Hoek, G., 2012. Variation of NO_2 and NO_x concentrations between and within 36 European study areas: Results from the ESCAPE study. *Atmos. Environ.* 62, 374–390.

[4] De Leeuw, F. a. a. M., Moussiopoulos, N., Sahm, P., Bartonova, A., 2001. Urban air quality in larger conurbations in the European Union. *Environ. Model. Softw.* 16, 399–414.

[5] Koop, G., McKitrick, R., Tole, L., 2010. Air pollution, economic activity and respiratory illness: Evidence from Canadian cities, 1974–1994. *Environ. Model. Softw.* 25, 873–885.

[6] Jephcote, C., Ropkins, K., Chen, H., 2014. The effect of socio-environmental mechanisms on deteriorating respiratory health across urban communities during childhood. *Appl. Geogr.* 51, 35–47.

[7] Koop, G., Tole, L., 2006. An investigation of thresholds in air pollution–mortality effects. *Environ. Model. Softw.* 21, 1662–1673.

[8] Gryparis, A., Dimakopoulou, K., Pedeli, X., Katsouyanni, K., 2014. Spatio-temporal semiparametric models for NO_2 and PM_{10} concentration levels in Athens, Greece. *Sci. Total Environ.* 479-480, 21–30.

[9] Oxley, T., Valiantis, M., Elshkaki, a., ApSimon, H. M., 2009. Background, road and urban transport modelling of air quality limit values (The BRUTAL model). *Environ. Model. Softw.* 24, 1036–1050.

[10] Namdeo, A., Mitchell, G., Dixon, R., 2002. TEMMS: an integrated package for modelling and mapping urban traffic emissions and air quality. *Environ. Model. Softw.* 17, 177–188.

[11] Carslaw, D. C., Ropkins, K., 2012. openair — An R package for air quality data analysis. *Environ. Model. Softw.* 27-28, 52–61.

[12] Gulliver, J., Briggs, D., 2011. STEMS-Air: a simple GIS-based air pollution dispersion model for city-wide exposure assessment. *Sci. Total Environ.* 409, 2419–29.

[13] Lin, M.-D., Lin, Y.-C., 2002. The application of GIS to air quality analysis in Taichung City, Taiwan, ROC. *Environ. Model. Softw.* 17, 11–19.

[14] Amann, M., Bertok, I., Borken-Kleefeld, J., Cofala, J., Heyes, C., Höglund-Isaksson, L., Klimont, Z., Nguyen, B., Posch, M., Rafaj, P., Sandler, R., Schöpp, W., Wagner, F., Winiwarter, W., 2011. Cost-effective control of air quality and greenhouse gases in Europe: Modeling and policy applications. *Environ. Model. Softw.* 26, 1489–1501.

[15] Chen, H., Namdeo, A., Bell, M., 2008. Classification of road traffic and roadside pollution concentrations for assessment of personal exposure. *Environ. Model. Softw.* 23, 282–287.

[16] Lefebvre, W., Degrawe, B., Beckx, C., Vanhulsel, M., Kochan, B., Bellemans, T., Janssens, D., Wets, G., Janssen, S., de Vlieger, I., Int Panis, L., Dhondt, S., 2013. Presentation and evaluation of an integrated model chain to respond to traffic- and health-related policy questions. *Environ. Model. Softw.* 40, 160–170.

[17] Eeftens, M., Tsai, M.-Y., Ampe, C., Anwander, B., Beelen, R., Bellander, T., Cesaroni, G., Cirach, M., Cyrys, J., de Hoogh, K., De Nazelle, A., de Vocht, F., Declercq, C., Dėdelė, A., Eriksen, K., Galassi, C., Gražulevičienė, R., Grivas, G., Heinrich, J., Hoffmann, B., Iakovides, M., Ineichen, A., Katsouyanni, K., Korek, M., Krämer, U., Kuhlbusch, T., Lanki, T., Madsen, C., Meliefste, K., Mölter, A., Mosler, G., Nieuwenhuijsen, M., Oldenwening, M., Pennanen, A., Probst-Hensch, N., Quass, U., Raaschou-Nielsen, O., Ranzi, A., Stephanou, E.,

Sugiri, D., Udvardy, O., Vaskövi, É., Weinmayr, G., Brunekreef, B., Hoek, G., 2012. Spatial variation of $PM_{2.5}$, PM_{10}, $PM_{2.5}$ absorbance and PMcoarse concentrations between and within 20 European study areas and the relationship with NO_2 – Results of the ESCAPE project. *Atmos. Environ.* 62, 303–317.

[18] Wang, M., Beelen, R., Stafoggia, M., Raaschou-Nielsen, O., Andersen, Z. J., Hoffmann, B., Fischer, P., Houthuijs, D., Nieuwenhuijsen, M., Weinmayr, G., Vineis, P., Xun, W. W., Dimakopoulou, K., Samoli, E., Laatikainen, T., Lanki, T., Turunen, A. W., Oftedal, B., Schwarze, P., Aamodt, G., Penell, J., De Faire, U., Korek, M., Leander, K., Pershagen, G., Pedersen, N. L., Östenson, C.-G., Fratiglioni, L., Eriksen, K. T., Sørensen, M., Tjønneland, A., Bueno-de-Mesquita, B., Eeftens, M., Bots, M. L., Meliefste, K., Krämer, U., Heinrich, J., Sugiri, D., Key, T., de Hoogh, K., Wolf, K., Peters, A., Cyrys, J., Jaensch, A., Concin, H., Nagel, G., Tsai, M.-Y., Phuleria, H., Ineichen, A., Künzli, N., Probst-Hensch, N., Schaffner, E., Vilier, A., Clavel-Chapelon, F., Declerq, C., Ricceri, F., Sacerdote, C., Marcon, A., Galassi, C., Migliore, E., Ranzi, A., Cesaroni, G., Badaloni, C., Forastiere, F., Katsoulis, M., Trichopoulou, A., Keuken, M., Jedynska, A., Kooter, I. M., Kukkonen, J., Sokhi, R. S., Brunekreef, B., Katsouyanni, K., Hoek, G., 2014. Long-term exposure to elemental constituents of particulate matter and cardiovascular mortality in 19 European cohorts: results from the ESCAPE and TRANSPHORM projects. *Environ. Int.* 66, 97–106.

[19] Oxley, T., ApSimon, H. M., 2007. Space, time and nesting Integrated Assessment Models. *Environ. Model. Softw.* 22, 1732–1749.

[20] Li, J., Heap, A. D., 2014. Spatial interpolation methods applied in the environmental sciences: A review. *Environ. Model. Softw.* 53, 173–189.

[21] Singh, V., Carnevale, C., Finzi, G., Pisoni, E., Volta, M., 2011. A cokriging based approach to reconstruct air pollution maps, processing measurement station concentrations and deterministic model simulations. *Environ. Model. Softw.* 26, 778–786.

[22] Robinson, D. P., Lloyd, C. D., McKinley, J. M., 2013. Increasing the accuracy of nitrogen dioxide (NO_2) pollution mapping using geographically weighted regression (GWR) and geostatistics. *Int. J. Appl. Earth Obs. Geoinf.* 21, 374–383.

[23] Fotheringham, A. S., Brunsdon, C., and Charlton, M. (2002). Geographically weighted regression: The analysis of spatially varying relationships. Chichester: Wiley.

[24] Johnston, K., Ver Hoef, J. M., Krivoruchko, K., and Lucas, N. (2001). Using ArcGIS Geostatistical Analyst. Redlands: ESRI Press.

[25] Matejicek, L., Engst, P., Janour, Z., 2006. A GIS-based approach to spatio-temporal analysis of environmental pollution in urban areas: A case study of Prague's environment extended by LIDAR data. *Ecol. Modell.* 199, 261–277.

[26] Prague City Hall, 2014. Yearbook - report on state of the environment, Prague Environment 2012. City of Prague. http://www.cenia.cz/ rocenka2011/index.htm. Last accessed in March 2016.

[27] CHMI, 2014. Air Pollution and Atmospheric Deposition in Data, the Czech Republic Annual Tabular Overview. Czech Hydrometeorological Institute, Air Quality Protection Division. http://portal.chmi.cz/files/ portal/docs/uoco/isko/tab_roc/tab_roc_EN.html. Last accessed in March 2016.

[28] IHIS, 2014. Health Statistics Yearbook of the Hl.m. Praha Region. Institute of Health Information and Statistics of the Czech Republic. http://www.uzis.cz/en/catalogue/health-statistics-yearbook-hlm-praha-region. Last accessed in March 2016.

[29] Robinson, D. P., Lloyd, C. D., McKinley, J. M., 2013. Increasing the accuracy of nitrogen dioxide (NO_2) pollution mapping using geographically weighted regression (GWR) and geostatistics. *Int. J. Appl. Earth Obs.* Geoinf. 21, 374–383. Robinson, D. P., Lloyd, C. D., McKinley, J. M., 2013. Increasing the accuracy of nitrogen dioxide (NO_2) pollution mapping using geographically weighted regression (GWR) and geostatistics. *Int. J. Appl. Earth Obs. Geoinf.* 21, 374–383.

[30] Zhang, L., Gove, J. H., Heath, L. S., 2005. Spatial residual analysis of six modeling techniques. *Ecol. Modell.* 186, 154–177.

[31] Li, S., Zhao, Z., Miaomiao, X., Wang, Y., 2010. Investigating spatial non-stationary and scale-dependent relationships between urban surface temperature and environmental factors using geographically weighted regression. *Environ. Model. Softw.* 25, 1789–1800.

[32] Ayres-Sampaio, D., Teodoro, A. C., Sillero, N., Santos, C., Fonseca, J., Freitas, A., 2014. An investigation of the environmental determinants of asthma hospitalizations: An applied spatial approach. *Appl. Geogr.* 47, 10–19.

[33] Roy, D. P., Wulder, M. a., Loveland, T. R., C. E., W., Allen, R. G., Anderson, M. C., Helder, D., Irons, J. R., Johnson, D. M., Kennedy, R., Scambos, T. a., Schaaf, C. B., Schott, J. R., Sheng, Y., Vermote, E. F., Belward, a. S., Bindschadler, R., Cohen, W. B., Gao, F., Hipple, J. D.,

Hostert, P., Huntington, J., Justice, C. O., Kilic, a., Kovalskyy, V., Lee, Z. P., Lymburner, L., Masek, J. G., McCorkel, J., Shuai, Y., Trezza, R., Vogelmann, J., Wynne, R. H., Zhu, Z., 2014. Landsat-8: Science and product vision for terrestrial global change research. *Remote Sens. Environ.* 145, 154–172.

[34] Anderson, H. R., 2009. Air pollution and mortality: A history. *Atmos. Environ.* 43, 142–152.

[35] Smith, T. W., Axon, C.J., Darton, R. C., 2013. The impact on human health of car-related air pollution in the UK, 1995–2005. *Atmos. Environ.* 77, 260–266.

[36] Pascal, M., Corso, M., Chanel, O., Declercq, C., Badaloni, C., Cesaroni, G., Henschel, S., Meister, K., Haluza, D., Martin-Olmedo, P., Medina, S., 2013. Assessing the public health impacts of urban air pollution in 25 European cities: results of the Aphekom project. *Sci. Total Environ.* 449, 390–400.

[37] Jephcote, C., Chen, H., 2013. Geospatial analysis of naturally occurring boundaries in road-transport emissions and children's respiratory health across a demographically diverse cityscape. *Soc. Sci. Med.* 82, 87–99.

[38] Portnov, B. a., Dubnov, J., Barchana, M., 2009. Studying the association between air pollution and lung cancer incidence in a large metropolitan area using a kernel density function. *Socioecon. Plann. Sci.* 43, 141–150.

In: Air Pollution
Editor: Gerald L. Burns

ISBN: 978-1-63485-374-3

Chapter 3

SHORT- AND LONG-TERM EFFECTS OF URBAN AIR POLLUTION ON CARDIOPULMONARY AND MALIGNANT DEATH RATES

Svetlana Stanišić Stojić[1,2*], Nemanja Stanišić[2] and Andreja Stojić[3]

[1]Faculty of Physical Chemistry, University of Belgrade, Belgrade, Serbia
[2]Singidunum University, Belgrade, Serbia
[3]Institute of Physics Belgrade, University of Belgrade, Belgrade, Serbia

ABSTRACT

Substantial scientific evidence suggests that exposure to even modest levels of air pollutants is associated with premature mortality. While relative contributions of pollutant species remain a matter of scientific debate, the current findings converge on the statement that short-term exposure over a period of few hours to weeks can aggravate the pre-existing health conditions, sometimes leading to a fatal outcome, whereas longer-term exposure reduces life expectancy by a few years. Susceptible groups including children and the elderly are at greater risk, as well as the people living in overpopulated low-income regions, due to being exposed to pollutant concentrations up to twenty times higher than those commonly observed in modern countries.

* Corresponding author: Email: sstanisic@singidunum.ac.rs.

The goal of this study is to present the findings on the effects of short- and long-term exposure to air pollution on death rates in Belgrade area (Serbia), based on the 6-year (2009-2014) regular pollutant monitoring and the corresponding administrative records on mortality. To that end, a quasi-Poisson regression model was combined with distributed lag non-linear model to examine this association after controlling for confounding variables such as temperature, seasonal trends and day of the week. The study discusses relative rates, expressed as % increase in circulatory, respiratory and malignant mortality per unit increase in daily average pollutant concentrations, together with the observed age- and gender-specific effects. According to the results, short-term exposure to elevated concentrations of PM_{10}, SO_2 and NO_2 is associated with relatively small mortality risk, whereas exposure during the last year appears to be more significant, particularly for circulatory mortality. Apart from the elderly group, severity of health effects is also observed for the male population under the age of 65. According to multiple economic indicators, Serbia is considered a developing country, even though its demographic trends are complementary with or less favourable than those typically anticipated in Western developed countries, notably due to high circulatory and malignant death rates. Poor air quality in Belgrade area is identified as an important environmental risk factor with pollutant concentrations exceeding the values observed in most European metropolis.

Keywords: air pollution, mortality, health effects, exposure

INTRODUCTION

The findings on toxicity and carcinogenicity of air pollutant species and their mixtures are consistent in epidemiological research, studies in experimental animals and studies of biogenic mechanisms (Jerrett et al., 2013). Ever since pre-industrial era (1860) to present days, significant changes in emissions, climate and methane concentrations have altered the atmospheric composition and led to an increase of surface ozone and $PM_{2.5}$ concentrations by 30 and 8 $\mu g\ m^{-3}$, respectively (Fang et al., 2013). The evidence from Greenland ice cores and Alpine glaciers confirmed the significant increase of sulfate and carbonaceous aerosol concentrations over the last 150 years (Fischer et al., 1998).

Current estimates of the global health burden indicate that between 3.3 and 3.7 million cardiopulmonary deaths per year can be associated with exposure to anthropogenic $PM_{2.5}$, with most cases observed in China and India

(Anenberg et al., 2010; Lelieveld et al., 2015). Lung cancer and ischemic heart disease are the two most represented causes, contributing with 12.8% and 9.4% to global premature mortality associated with $PM_{2.5}$, respectively (Evans et al., 2013). In addition to particulate matter, adverse global health effects of ozone were also estimated using satellite-based observations and extensive surface measurements. Brauer et al., (2012) concluded that between 50,000 and 270,000 premature respiratory deaths were associated globally and annually with exposure to anthropogenic ozone, whereas the findings of Silva et al., (2013) suggested that ozone-related respiratory mortality might be even higher, with up to 470,000 deaths per year. Notwithstanding the fact that the levels of particulate matter and inorganic gaseous oxides in developed countries have declined significantly over the last few decades, reducing ozone levels might be a challenge, as it is generated through a complex set of atmospheric photochemical reactions of primary pollutants.

Major emission sources in modern countries include fossil fuel combustion for industrial needs, traffic and power generation, whereas in developing countries of Asia, residential biomass burning for heating and cooking is still a matter of concern. The observed daily $PM_{2.5}$ concentrations in mega-cities of Asia, ranging from 100 to 300 $\mu g\ m^{-3}$, are up to twenty times higher than those commonly observed in developed countries (Shah et al., 2013). For instance, the annual average $PM_{2.5}$ concentrations of 75 $\mu g\ m^{-3}$ in the capital city of Mongolia are mostly driven by extremely high winter concentrations (250 $\mu g\ m^{-3}$) due to coal and wood combustion in low-income traditional housing areas. These emissions are estimated to cause 29% of cardiopulmonary and 40% of lung cancer mortality in Ulaanbaatar, corresponding to 10% of total non-accidental death rates (Allen et al., 2013). Unfortunately, with the exception of China, the air pollution studies in Asia and developing countries worldwide are relatively scarce due to poor environmental monitoring and a paucity of mortality surveillance data. In Europe, Russia and eastern parts of the USA and Asia, agricultural sources significantly contribute to air quality deterioration (Lelieveld et al., 2015). In addition to anthropogenic pollution prevailing in urban areas, the impact of natural particles, such as mineral dust, on premature mortality should not be underestimated. According to Giannadaki et al., (2014), desert dust is associated with up to 50% of cardiopulmonary deaths in the population from twenty countries across the so-called 'dust belt' from North Africa to East Asia.

Brauer et al., (2012) reported that almost 90% of the world's population in 2005 lived in the areas with the annual mean $PM_{2.5}$ concentrations exceeding

10 μg m^{-3}, as recommended by the WHO Air Quality Guidelines. Thereby, children under the age of 5 living in low-income communities are considered the most susceptible group due to their poor nutritional status and the fact that their families lack the knowledge and resources necessary to avoid exposure. China was among the first developing countries to adopt large scale regulations on power plant emissions, which resulted in 20% reduction of infant death rates (Tanaka, 2015). The emissions related to biomass combustion also threaten public health in rural areas of developed countries. Noonan et al., (2011) reported that the single wood stove intervention in a rural area of the USA resulted in 27.6% lower $PM_{2.5}$ winter concentrations and the corresponding reduction in respiratory infections among school children. The findings of Heinrich et al., (2002) also suggested that the decreasing prevalence of non-allergic respiratory disorders in children was associated with a decline in total suspended particles and SO_2 in the eastern part of Germany after unification.

According to some projections, the number of deaths caused by exposure to air pollution could double by 2050. The fact that the exposure dose-response relationship was found to be linear down to $PM_{2.5}$ concentrations of 8 μg m^{-3} (Lepeule et al., 2012) and that modest enhancements in air quality would have health benefits, has prompted further rethinking of the current air quality guidelines. In terms of economic costs, reduction of 3.9 μg m^{-3} in mean $PM_{2.5}$ concentrations would prevent almost 8,000 cardiovascular hospitalizations and save one third of billion US dollars annually (Shah et al., 2013). Unfortunately, there is still a prevailing concern in developed countries that the financial expenditures resulting from air pollution abatement measures might outweigh the potential health benefits.

In this study, we present the findings regarding the association between PM_{10}, NO_2 and SO_2 exposure and mortality caused by circulatory, respiratory and malignant diseases in Belgrade, Serbia, based on 2009-2014 dataset. Besides the fact that PM_{10} concentrations in Belgrade area are higher than concentration measurements in other European cities (Perišić et al., 2014; Stojić et al., 2015a; Stojić et al., 2015b; Stojić et al., 2015c; Todorović et al., 2015; Šoštarić et al., 2015), recent studies have suggested that suspended particles contain high concentrations of pollutant matter, arsenic and benzo(a)pyrene, resulting from fossil-fuel combustion during winter months (Stojić et al., 2016). In order to examine both short- and long-term effects of air pollution, we used the moving average pollutant concentrations during two days (lag 01) and a year (lag Y) preceding death. The effects of air pollution on mortality follow a complex pattern across different geographic areas, depending on the population health characteristics and their technological and economic capacity to adapt, as well as due to differences in exposure level, particle composition and methodological approach. The studies that have limited their analyses to acute health effects experienced within few days upon exposure mainly emphasize the increase in death counts and hospitalizations during and immediately after pollution episodes. However, the question remains as to whether there is a critical exposure time window that is primarily responsible for the increase in death rates. The overall evidence suggests that the mortality estimates increase in magnitude as the follow-up period increases (Goodman et al., 2004).

MATERIALS AND METHODS

Mortality statistics, including total (cause-of-death coding A00-R99), circulatory (I00-I99), respiratory (J00-J99) and malignant (C00-D48) mortality, were retrieved from the Institute of Public Health Belgrade (IPHB), for the period from 2009-2014. The causes of death were coded according to the International Classification of Diseases, 10th Revision, ICD-10, excluding the deaths attributed to injuries, poisoning and external causes from the overall death rate. The concentrations of PM_{10}, NO_2 and SO_2 were obtained from 7, 21 and 22 monitoring stations within the IPHB network, respectively, uniformly distributed throughout the city area. Measurements were conducted at automatic stations using beta-ray attenuation samplers (Thermo FH 62 I-R) for PM_{10} and referent sampling devices, Horiba APNA 360 and APSA 360 analyzers, for NO_2 and SO_2, respectively. At semi-automatic measuring stations, PM_{10} concentrations were determined using referent Sven Leckel samplers, whereas determination of NO_2 and SO_2 concentrations was performed based on ISO 6768:1998 (Modified Griess-Saltzman method) and ISO 6767:1990 (tetrachloromercurate (TCM)/pararosaniline method) standards, respectively. Temperature data were obtained from the Global Data Assimilation System with the spatial resolution of 1 degree.

As single-day models were reported to underestimate the cumulative effects of air pollutants on mortality (Meng et al., 2012), we specified a model which, besides temperature and other factors affecting mortality such as season, temperature and day of the week, included three categories of air pollutants: PM_{10}, NO_2 and SO_2.

The effect of temperature is shown to be both delayed in time and shifted by the so-called 'harvesting effect' – a temporal advance in deaths that would occur later in time in the absence of exposure to heat or cold (Ye et al., 2012). The harvesting effect is typically followed by a period of reduced mortality. In order to simultaneously capture the non-linear exposure-response dependencies and delayed temperature effects, we used distributed lag non-linear models (DLNM). This methodology is based on the definition of 'cross-basis,' a bi-dimensional space of functions that describes simultaneously the shape of the relationship along both the space of the predictor and the lag dimension of its occurrence (Gasparrini et al., 2010). The implementation of DLNM modeling framework (Gasparrini, 2011) is available in the statistical environment R (R Core Team, 2014). Based on the methodology presented in Gasparrini et al., (2015), we modeled the exposure-response curve with a quadratic B-spline with three internal knots placed at the 10th, 75th and 90th percentiles of the observed local temperature distribution and the lag-response curve (maximum lag up to 21 day) with a natural cubic B-spline with an intercept and three internal knots placed at equally-spaced values in the log scale. In order to control for seasonal and long-term effects, we included a natural cubic B-spline of time with 8 degrees of freedom per year. The model also included an indicator of day of the week. Finally, the effects of pollutants are included as simple moving average of their observed concentration level in 10 μg m^3 over 2 (commonly referred in the literature as lag 01) and 366 (lag Y) consecutive days ending on, and inclusive of day *t*.

Accordingly, the quasi-Poisson[1] regression model is specified as follows:

$$E(Yt) = exp\{\alpha + \beta T_{t,l} + NS(time,df) + \lambda DOW_t + PM_{10t,01} + PM_{10t,Y} + O_{2t,01} + NO_{2t,Y} + SO_{2t,01} + SO_{2t,Y}\}$$

where Yt is a daily death count on day t; α is the intercept; $T_{t,l}$ is a matrix of variables obtained by the transformation of temperature, β is a vector of coefficients for $T_{t,l}$ and l is the lag day; *NS(time, df)* is the natural cubic spline of time; *DOW* and λ are dummy variables representing day of the week and the corresponding vector of coefficients, respectively.

Weekly, seasonal and annual dynamics, and trend (Pretty, 2015) of the mortality and air pollutant concentrations were analyzed using the Openair (Carslaw and Ropkins, 2012) package within the statistical software environment R (Team, 2012).

[1] Quasi-Poisson is used in order to compensate for overdispersion (i.e., residual deviance is larger than residual degrees of freedom).

Results and Discussion

Seasonal Variations in Mortality and Air Pollution

During the period from 2009-2014, there were 113,430 deaths registered in Belgrade area, of which 60,407 cases (i.e., 53.25%) were attributed to circulatory diseases, 4,649 cases (i.e., 4.01%) were associated with respiratory failure, while 31,191 cases (i.e., 27.50%) were caused by malignant diseases. Annual circulatory mortality in Belgrade area over the last 15 years has been ranging between 635 to 677 per 100,000 inhabitants (Statistical Office of the Republic of Serbia, 2015). When compared with the death rates observed during the same period in Europe (Detailed Mortality Database, 2015), these values are among the highest ranking. Annual death rates caused by malignant diseases have also been extremely high over the past 15 years, ranging from 260 to 323 cases per 100,000 inhabitants. On the other hand, annual respiratory mortality, ranging from 28 to 60 cases per 100,000 inhabitants, can be considered normal in relation to the figures observed in other European countries. As regards deseasonalised trend, circulatory mortality shows stagnation (0.36%) during the observed period, whereas malignant (1.94%) and respiratory mortality (6.06%) record moderate, i.e., significant growth rates (Figure 1).

Mortality caused by circulatory and respiratory diseases exhibits a strong seasonal pattern, with highest death rates registered in February and March and minimum occurrence during summer months (Figure 2). Whereas mortality rates of the elderly population also indicate extreme seasonal variations, mortality rates of the population under the age of 65 exhibit different distribution patterns throughout the year, probably due to the effects of numerous other factors apart from air pollution. Likewise, mortality caused by malignant diseases exhibits no seasonal variations, with more than one third of deaths due to malignant diseases observed among the population younger than 65 years, which indicates poor health of Belgrade population. Unlike that, the data obtained from the Cancer Research Center in the UK for the period from 2010-2012 showed that up to 78% of such diseases were observed in the population aged 65 years and over. The health status of the population in Serbia correlates with the anticipated life expectancy. According to the Statistical Office of the Republic of Serbia (2015), it was estimated at 74.3 years, which is considerably lower than the EU-28 average value of 80.6 years (Mortality and life expectancy statistics, 2013).

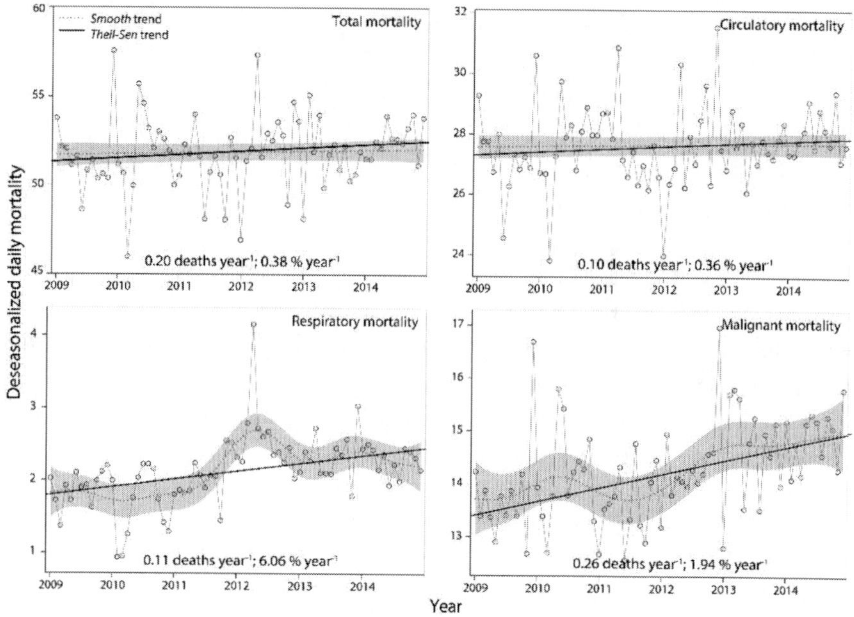

Figure 1. Mortality trend analysis in the period from 2009 to 2014.

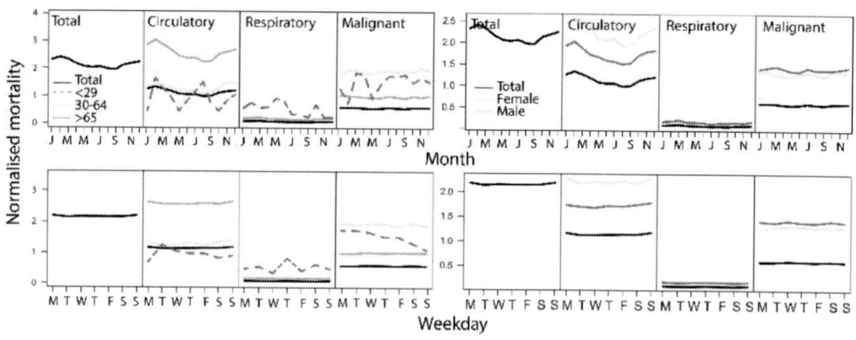

Figure 2. Mortality seasonal and weekday variations by age (left) and gender (right) in the period from 2009 to 2014.

The statistics showed that circulatory, respiratory and malignant mortality differ considerably in terms of gender structure: mortality caused by respiratory and malignant diseases was more common in male population, while circulatory mortality was prevalent in female population (Table 1).

Table 1. Mortality statistics in the period from 2009 to 2014

Mortality	Gender	Age	Mean	Sum	Min.	Max.
Total			51.84	113430	25	83
Circulatory	Total		27.61	60407	10	49
	Female	Total	3.75	32847	0	31
		<29	0.02	42	0	1
		30-64	1.02	2227	0	6
		>65	13.97	30572	3	31
		Unknown	0.00	6	0	1
	Male	Total	3.15	27560	0	22
		<29	0.02	54	0	1
		30-64	2.27	4975	0	8
		>65	10.29	22517	0	22
		Unknown	0.01	14	0	1
Respiratory	Total		2.12	4649	0	9
	Female	Total	0.23	1979	0	5
		<29	0.01	26	0	1
		30-64	0.14	301	0	3
		>65	0.75	1651	0	5
		Unknown	0.00	1	0	1
	Male	Total	0.31	2670	0	6
		<29	0.01	28	0	1
		30-64	0.21	459	0	3
		>65	1.00	2180	0	6
		Unknown	0.00	3	0	1
Malignant	Total		14.26	31191	3	29
	Female	Total	2.19	14395	0	14
		<29	0.03	64	0	2
		30-64	2.23	4886	0	9
		>65	4.32	9445	0	14
	Male	Total	2.56	16796	0	14
		<29	0.04	89	0	2
		30-64	2.61	5720	0	11
		>65	5.02	10987	0	14

Zhang et al., (2012) elaborate on the higher risk for mortality in women hospitalized with an acute myocardial infarction versus their male counterparts, relying mainly on the fact that women suffer from greater number of chronic diseases at the time of a heart attack, thus making them more susceptible to fatal outcome. Namely, previous research has shown that women were more prone to non-fatal chronic diseases throughout life, as well as to gastrointestinal and respiratory infections. Unlike them, men were generally healthier throughout life. However, their health deteriorates with age, especially vital organs (coronary and malignant diseases, emphysema, cirrhosis of the liver and chronic kidney disease), thus supporting the statement that life expectancy in men is few years shorter (Ross, Masters & Hummer, 2012). However, Salam et al., (2013) showed that, after adjusting for age, diabetes mellitus, and acute coronary syndrome, female gender remained a factor independently related to increased in-hospital mortality, thus underpinning the claim that there are other factors that could contribute significantly to greater mortality rates caused by cardiovascular diseases in women, besides age and chronic diseases.

The daily mean concentrations of PM_{10}, SO_2 and NO_2 for the entire period were 48.31, 32.68 and 15.52 $\mu g\ m^{-3}$, respectively. The levels of PM_{10} and SO_2 exhibited a clear seasonal pattern, with highest concentrations observed during the cold season (Figures 3 and 4). The emissions of SO_2 are predominantly associated with sulfur-rich solid fuel-burning in central heating plants and industrial facilities, whereas PM_{10} originates from a number of sources and processes, including atmospheric photochemical reactions.

As shown in Table 2, the moving average PM_{10} and SO_2 concentrations exhibit a high correlation, thus suggesting that their variations, particularly those long-term, are driven by the same anthropogenic and meteorological factors. Conversely, NO_2 concentrations exhibited pronounced weekly variations, whereas the average NO_2 values for the cold and warm season were almost identical (Stanišić Stojić et al., 2015). This could be partly explained by the fact that NO_2 levels are considered an indicator of traffic-related emissions that could be observed throughout the year.

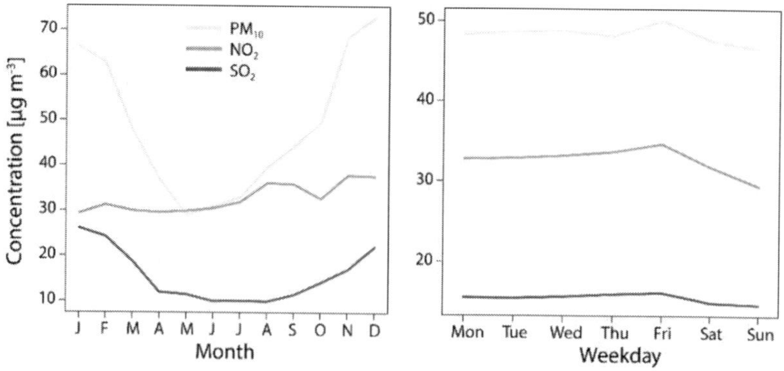

Figure 3. Pollutant seasonal and weekday variations.

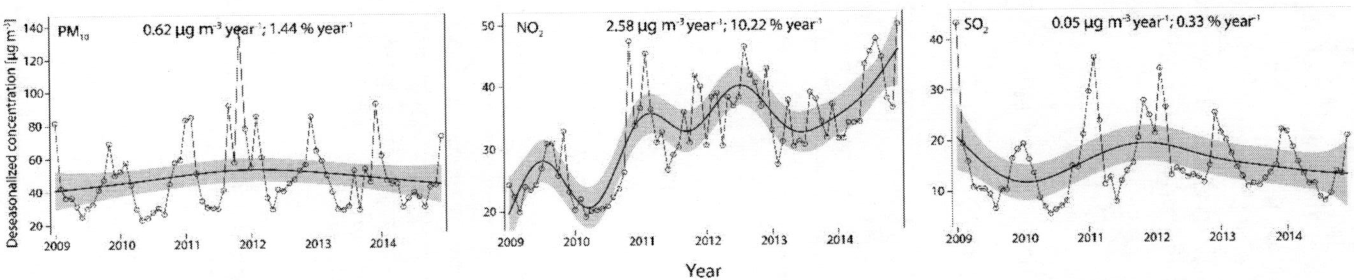

Figure 4. Pollutant trend analysis.

As can be seen in Table 3, the number of days with average PM_{10} concentrations exceeding 50 μg m^{-3} was in the range from 75 to 155 per year, which is considerably higher than the margin of tolerance set by the Air Quality Standard (35 exceedances per year). However, the mean annual NO_2 concentrations did not exceed the target value of 40 μg m^{-3}, nor any mean daily SO_2 concentrations higher than the recommended limit of 135 μg m^{-3} could be observed.

Table 2. Correlations between pollutant species

	PM_{10} 01	PM_{10} Y	NO_2 01	NO_2 Y	SO_2 01	SO_2 Y
PM_{10} 01	1	0.15	0.51	0.08	0.71	0.18
PM_{10} Y		1	0.26	0.64	0.18	0.96
NO_2 01			1	0.37	0.43	0.24
NO_2 Y				1	0.05	0.68
SO_2 01					1	0.21
SO_2 Y						1

Table 3. Number of days with PM_{10} concentration exceeding 50 μg m^{-3} and mean annual concentrations of PM_{10}, NO_2 and SO_2 [μg m^{-3}] in the period from 2009 to 2014

Year	Number of days with PM_{10} exceedances	PM_{10}	NO_2	SO_2
2009	81	43.63	25.78	15.22
2010	75	39.73	24.69	11.93
2011	155	62.37	34.64	19.59
2012	150	52.66	38.33	17.71
2013	116	47.80	33.44	15.06
2014	90	43.66	39.22	13.60

Short-Term and Long-Term Effects of Air Pollution on Mortality

The data examined reveals that the average daily number of circulatory deaths was higher during the 15 days with highest pollutant concentrations (Table 4). Furthermore, the increase in total mortality was mostly driven by the increase in circulatory deaths, as the most prevailing type. The effects of pollution on respiratory and malignant mortality are not quite clear, probably due to long evolution and variety of malignant diseases, i.e., relatively small number of observations related to respiratory diseases and related mortality.

Table 4. Average daily number of deaths over the entire period (2009-2014) and average daily number of deaths during 15 days with highest pollutant concentrations

	Entire period	PM_{10}	NO_2	SO_2
Total	51.8	53.8	54.9	59.6
Circulatory	27.6	30.1	29.4	36.0
Respiratory	2.1	1.9	1.7	2.4
Malignant	14.2	13.9	15.5	13.1

Table 5. The estimated percentage changes in circulatory mortality due to a 10 μg m^{-3} increase in moving average pollutant concentrations during two days (lag 01) and a year (lag Y) preceding death

	<65	>65	Male	Female
PM_{10} 01	-0.001	0.074***	0.086**	0.047
	(-0.140, 0.100)	(0.019, 0.120)	(0.013, 0.150)	(-0.024, 0.100)
PM_{10} Y	0.632	0.279	-3.021	3.228*
	(-6.593, 8.329)	(-2.634, 3.252)	(-6.735, 0.803)	(-0.584, 7.144)
NO_2 01	-0.195	-0.243***	-0.268**	-0.209*
	(-0.606, 0.200)	(-0.407, 0.000)	(-0.484, 0.000)	(-0.417, 0.000)
NO_2 Y	12.708	7.824*	15.599***	2.672
	(-7.725, 37.575)	(-0.414, 16.649)	(4.057, 28.415)	(-7.235, 13.542)
SO_2 01	0.645*	0.096	0.411**	-0.052
	(-0.065, 1.308)	(-0.188, 0.300)	(0.037, 0.783)	(-0.416, 0.300)
SO_2 Y	0.763	-3.308	12.649	-14.109
	(-32.841, 51.135)	(-17.336, 13.088)	(-8.610, 38.819)	(-29.674, 4.812)
Tuesday	-11.928***	-1.947	-2.041	-4.131*
	(-19.832, -2.955)	(-5.456, 1.613)	(-6.703, 2.840)	(-8.472, 0.401)
Wednesday	-6.846	-2.962	-3.040	-3.799
	(-15.098, 2.122)	(-6.454, 0.602)	(-7.677, 1.816)	(-8.163, 0.702)
Thursday	-4.234	-3.262*	-1.830	-4.671**
	(-12.674, 4.917)	(-6.752, 0.300)	(-6.518, 3.045)	(-9.016, 0.000)
Friday	-10.374**	-0.950	-2.541	-1.762
	(-18.417, -0.995)	(-4.510, 2.737)	(-7.210, 2.327)	(-6.217, 2.840)
Saturday	0.296	-2.719	0.820	-4.947**
	(-8.439, 9.856)	(-6.215, 0.904)	(-3.950, 5.760)	(-9.273, 0.000)
Sunday	4.057	0.304	2.365	-0.542
	(-4.884, 13.769)	(-3.249, 3.977)	(-2.435, 7.358)	(-4.984, 4.081)

Table 6. The estimated percentage changes in malignant mortality due to a 10 μg m^{-3} increase in moving average pollutant concentrations during two days (lag 01) and a year (lag Y) preceding death

	<65	>65	Male	Female
PM_{10} 01	-0.126**	0.009	-0.001	-0.081
	(-0.248, 0.000)	(-0.076, 0.000)	(-0.098, 0.000)	(-0.184, 0.000)
PM_{10} Y	2.534	0.998	0.351	2.963
	(-3.848, 9.308)	(-3.495, 5.654)	(-4.702, 5.654)	(-2.495, 8.654)
NO_2 01	0.219	0.129	0.108	0.220
	(-0.141, 0.501)	(-0.122, 0.300)	(-0.179, 0.300)	(-0.083, 0.501)
NO_2 Y	-1.395	8.436	10.928	-2.221
	(-17.319, 17.586)	(-4.091, 22.507)	(-3.598, 27.634)	(-15.622, 13.202)
SO_2 01	0.345	0.413*	0.195	0.622**
	(-0.294, 0.904)	(-0.038, 0.803)	(-0.318, 0.702)	(0.085, 1.157)
SO_2 Y	-18.386	0.816	4.857	-16.889
	(-42.616, 16.067)	(-21.110, 28.788)	(-20.605, 38.403)	(-38.275, 11.851)
Tuesday	1.648	-0.540	-1.669	2.492
	(-6.094, 9.966)	(-5.969, 5.127)	(-7.726, 4.707)	(-4.184, 9.527)
Wednesday	1.702	3.434	2.507	3.255
	(-6.056, 10.076)	(-2.167, 9.308)	(-3.751, 9.090)	(-3.470, 10.407)
Thursday	-0.545	-0.731	-2.662	1.745
	(-8.184, 7.681)	(-6.167, 4.917)	(-8.685, 3.666)	(-4.916, 8.872)
Friday	-5.197	-0.190	-3.629	0.105
	(-12.574, 2.737)	(-5.651, 5.548)	(-9.613, 2.737)	(-6.477, 7.144)
Saturday	0.599	-1.004	-1.289	0.550
	(-7.092, 8.872)	(-6.411, 4.707)	(-7.363, 5.127)	(-6.031, 7.573)
Sunday	-3.791	0.000	-1.573	-1.011
	(-11.200, 4.185)	(-5.429, 5.654)	(-7.611, 4.812)	(-7.491, 5.866)

According to the model, no substantial change was observed in mortality immediately after air pollution episodes. Thus, a 10 μg m^{-3} increase in the average pollutant concentration is estimated to cause a negligible increase in death counts, rarely exceeding 0.5%, with statistically significant associations between PM_{10} and SO_2 and circulatory mortality in elderly and males, and SO_2 and malignant mortality in elderly and female population (Table 5 and 6). Low to moderate short-term health effects of air pollution were also reported in the extensive review of Brook et al., (2010). They concluded that the elevated $PM_{2.5}$ concentrations over 5 days led to the increase in circulatory mortality risk by 0.4% to 1.0%, even though the burden was not equally distributed within the population. Thereby, it should be taken into account that the health effects of fine particles are more severe compared to PM_{10}, because smaller diameter particles are more likely to penetrate deeply in the tracheobronchial and alveolar regions. The systematic review and meta-analysis of Shah et al.,

(2013) confirmed that heart failure hospital admissions and mortality were associated with moderate short-term increases in concentrations of particulate matter and all inorganic gaseous oxides, except for ozone.

In contrast to this, the meta-analysis including 33 time-series and case-crossover studies conducted in China, suggested that a 10 μg m^{-3} reduction in $PM_{2.5}$ concentrations would result in a considerable decrease (1.7% to 6.2%) in all-cause mortality (Shang et al., 2013). This heterogeneity in effect estimates across studies might be attributed to a number of factors, including pollutant characteristics and severity of pollution episodes in the study area. The complexity of the relationship between mortality and air pollution is driven by the fact that risk estimates can vary significantly across regions with similar characteristics, as shown in the spatial analysis for Los Angeles and New York incorporating data for 1.2 million adults from 172 U.S. urban areas (Krewski et al., 2009).

In the study aimed at investigating circulatory health effects of inorganic gaseous oxides, particulate matter and ozone, Milojević et al., (2014) used lags of 1 to 4 days and found no clear relationship between circulatory-related mortality, hospital admissions and air pollution, with the exception of $PM_{2.5}$ that was associated with arrhythmias, atrial fibrillation and pulmonary embolism, and NO_2 associated with an increased risk of myocardial infarction. Likewise, in the Danish cohort study of Raaschou-Nielsen et al., (2012), mean levels of NO_2 at the residence were also significantly associated with circulatory mortality after adjustment for potential confounders. As shown in Table 5 and Table 6, short-term exposure to NO_2 exhibited negative associations with death rates caused by circulatory and malignant diseases, which could indicate that the exposure to high NO_2 levels reaches its maximum effect upon the examined two-day period (lag 01). This assumption seems more probable when taking into account that the long-term exposure to NO_2 (lag Y) is associated with significant increase of circulatory and malignant mortality in people older than 65 years, particularly male population.

As regards long-term effects of air pollution on mortality, the herein presented results comply with the findings of Lepeule et al., (2012), suggesting that the exposure over longer period preceding death might be more significant than short-term exposure.

Thereby, the effects of PM_{10} were estimated as less significant compared to the effects of gaseous oxides (Tables 5, 6 and 7). The statistically significant associations were observed between circulatory mortality and exposure to NO_2 in elderly, particularly male population, as well as between circulatory mortality and exposure to PM_{10} in females. Similarly, the study of Chen et al., (2013), including data for three cities in Canada, suggested that long-term exposure to NO_2 was associated with a significant increase in circulatory mortality. Furthermore, Brook et al., (2010) reported a strong association between NO_2 and non-accidental mortality across 10 cities in Canada. Although these estimations cannot be ruled out, the authors assumed that NO_2 concentrations were acting as an indicator of some other variable or process. For instance, NO_2 levels exhibit high correlations in time and space with a range of toxic traffic-related species, including volatile organic compounds and polycyclic aromatic hydrocarbons, which could also contribute to the observed effects.

There isn't currently enough evidence to support the relationship between long-term exposure to air pollution and non-malignant respiratory mortality, although studies aimed at investigating short-term effects suggested that mid-term lags exceeding 40 days appear to be significant (Dimakopoulou et al., 2014). That is chiefly due to the fact that smaller incidence of respiratory mortality contributed to larger confidence intervals and larger variability of the effect estimates across studies (Hoek et al., 2013). Carey et al., (2013) investigated the relationship between individual annual exposure to air pollution and mortality in an English cohort and found a higher relative risk for respiratory deaths, which differed from most of the U.S. studies where no significant associations have been identified between particulate matter and non-malignant respiratory mortality. However, it is important to note that respiratory mortality rates in the UK were among the highest in Europe, thus leaving it an open question whether the conclusion proposed by Carey et al., (2013) can be generalized (Detailed Mortality Database, 2015).

It is reasonable to assume that the exposure-response time depends on the doze, physical and chemical properties of air pollutants, as well as the characteristics of the exposed population. Therefore, determining a unique time frame for assessing the impact of all pollutant matter might not lead to the anticipated results. As can be seen, the effects of long-term exposure to PM_{10} and SO_2 in most of the cases have the reverse effects in female and male population, which may indicate that the examined exposure time window, in this case a year preceding death, corresponded to the maximum effect in one gender group and post-harvesting effect in the other group. As previously mentioned, the post-harvesting reduction in death rates is expected to follow and balance the so-called 'harvesting effect' i.e., temporally advanced deaths of fragile population. In other words, it is possible that the effects of long-term exposure to SO_2 on circulatory mortality occurred first in females, while the reverse scenario applies for respiratory mortality.

Table 7. The estimated percentage changes in respiratory mortality due to a 10 μg m^{-3} increase in moving average pollutant concentrations during two days (lag 01) and a year (lag Y) preceding death

	<65	**>65**	**Male**	**Female**
PM_{10} 01	0.302	0.059	0.100	0.126
	(-0.088, 0.602)	(-0.140, 0.200)	(-0.141, 0.300)	(-0.152, 0.401)
PM_{10} Y	-2.478	-1.179	1.291	-4.278
	(-22.107, 22.018)	(-11.058, 9.746)	(-11.093, 15.373)	(-17.397, 10.849)
NO_2 01	-0.508	-0.587*	-0.453	-0.763*
	(-1.703, 0.702)	(-1.184, 0.000)	(-1.186, 0.200)	(-1.583, 0.000)
NO_2 Y	-24.294	10.727	12.785	-6.897
	(-57.938, 36.206)	(-16.360, 46.521)	(-20.223, 59.361)	(-37.108, 37.713)
SO_2 01	0.891	-0.743	-0.394	-0.467
	(-1.016, 2.737)	(-1.777, 0.300)	(-1.635, 0.803)	(-1.877, 0.904)
SO_2 Y	94.657	-15.108	-21.260	25.487
	(-40.400, 535.346)	(-51.091, 47.256)	(-60.276, 56.049)	(-41.886, 170.743)
Tuesday	-10.022	-2.139	-5.886	-0.074
	(-32.125, 19.244)	(-14.044, 11.405)	(-20.034, 10.738)	(-16.517, 19.602)
Wednesday	-10.206	-3.158	-2.238	-7.137
	(-32.277, 19.006)	(-15.014, 10.296)	(-16.852, 14.912)	(-22.724, 11.516)
Thursday	-5.439	10.014	4.007	11.761
	(-28.468, 24.982)	(-3.095, 24.857)	(-11.367, 22.018)	(-6.252, 33.109)
Friday	2.946	-4.581	0.974	-8.859
	(-21.733, 35.391)	(-16.365, 8.763)	(-14.103, 18.649)	(-24.271, 9.636)
Saturday	-10.364	-7.547	-7.624	-8.560
	(-32.425, 18.887)	(-18.979, 5.443)	(-21.617, 8.763)	(-23.932, 9.856)
Sunday	-5.048	8.611	1.462	12.723
	(-28.011, 25.232)	(-4.189, 22.998)	(-13.448, 18.887)	(-5.167, 33.910)

The current findings of gender-specific effects of air pollution on mortality are inconsistent. Chen et al., (2012) reported larger effect estimate of PM_{10} on total mortality in females than in males, which can be explained by lower smoking rate in females. In epidemiological studies, smoking is a potential confounding factor because it plays an important role in numerous pathological conditions, including malignant, lung and cardiovascular diseases. The interpretation of Chen et al., (2012) is based on the assumption that the smoking effects dominate to the extent that the exposure to air pollution may not further exacerbate the health condition. The findings of Turner et al., (2011), who examined the relationship between long-term $PM_{2.5}$ concentrations and lung cancer mortality among 188,699 lifelong never smokers, suggested that a 10 μg m^{-3} increase in mean $PM_{2.5}$ concentrations was associated with a 15-27% increase in lung cancer death rates. Conversely, the study of Pope et al., (2004) demonstrated that larger mortality risk associated with air $PM_{2.5}$ exposure is observed in smokers relative to never smokers, which they explained by additive if not synergistic impact of pollutants from smoke and ambient air.

As summarized by Solomon et al., (2012), gender might be the factor that influences susceptibility, due to the size of airways, the dose received and related particle clearance. Furthermore, it could be assumed that exposure to air pollution can be higher in male population as certain outdoor manual occupations, particularly in traditional surroundings, are more common in men, which is why more significant effects of air pollution can be expected in male population.

Aside from gender-specific health effects of air pollution, some uncertainties still remain regarding the pathophysiological mechanism of pollution-related mortality. Even though epidemiological studies are not designed to examine mechanisms, the observed time course of response might provide some information (Pope et al., 2004). For instance, it takes at least 24 hours for acute inflammatory response to develop, whereas neural response, i.e., disturbance of autonomic neural system, might be the underlying cause of myocardial infarction and vasoconstriction immediately upon exposure (Peters, 2005). During the long-term exposure, both mechanisms are present together with oxidative stress-induced DNA damage (Bräuner et al., 2007), changes in the expression of genes involved in DNA damage and repair (DeMarini, 2013) and changes in blood coagulability (Seaton et al., 1995). According to Milojević et al., (2014), the lower risk of ST elevation myocardial infarction and cerebrovascular insult and higher prevalence of non-myocardial infarction outcomes suggests that pollution-related circulatory health effects were probably more mediated by non-thrombotic pathways.

Evidence from 'Natural Experiments'

A limited number of studies investigated shifts in mortality associated with stringent environmental policies, and provided an important insight into substantial improvements in air quality and related health benefits. The study of Correia et al., (2013), based on the data from 545 USA counties, showed that a decrease in the annual average $PM_{2.5}$ concentrations over the 7-year period was associated with an increase in the mean life expectancy of 0.35 years, with the effect being pronounced in urban and densely populated areas.

The study of Clancy et al., (2002) showed that the effects of coal burning regulations in Dublin resulted in 16% and 64% declines in wintertime concentrations of SO_2 and black smoke, respectively. Accordingly, non-accidental, respiratory and circulatory deaths decreased by 5.7%, 15.5% and 10.3%, respectively. Similarly, a significant decrease of SO_2 concentrations occurred after banning sulfur-rich fuel in Hong Kong. The restrictions led to a decrease in risk for respiratory (3.9%) and circulatory (2.0%) mortality, whereas the average life expectancy increased by 20 and 41 days for females and males, respectively (Medley et al., 2002). The study conducted in Switzerland also supports the claim that long-term improvements in air quality and decreasing PM levels have beneficial effects on respiratory symptoms, including chronic cough, wheezing and breathlessness among adults (Schindler et al., 2009).

The sudden air quality deterioration caused by natural disasters or anthropogenic activities is another example of the so-called 'natural experiment' that eases concerns over confounding and exposure misclassification (Parker et al., 2008). The study of Shaposhnikov et al., (2014) investigated the combined effects of heat wave and air pollution (PM_{10} and ozone) on daily variations in death rates during wildfire in Moscow, Russia. According to the findings, with daily mean PM_{10} levels exceeding 300 $\mu g\ m^{-3}$, a steep increase in mortality risk was observed during the first two weeks, particularly for deaths related to nervous system, genitourinary and cerebrovascular diseases. The study of Chen et al., (2013) investigated the effects of China's Huai River policy that provided free coal for heating boilers in northern China cities. They showed that concentrations of total suspended particles were about 55% higher, whereas the life expectancy was about 5.5 years shorter in the North, due to the increased incidence of both circulatory and respiratory diseases.

Limitations of Epidemiological Studies

Some general limitations of epidemiological studies aimed at health risk estimation should be acknowledged. By analyzing the effects of environmental stressors, the entire population is considered equally exposed and the results obtained from a number of monitoring stations are assumed to be representative of individual exposure. More recent studies also used land regression models to derive individualized residential exposure (Jerrett et al., 2013). However, this is not realistic due to the fact that people exhibit different activity patterns and spend a significant period of time indoors (Basu and Samet, 2002). Furthermore, spatial distribution of particles and gaseous pollutants differs considerably: the concentrations of SO_2 and NO_2 are more closely related to their sources compared to PM_{10} concentrations, which are found to be relatively homogeneous over a larger area (Beelen et al., 2014). Rural areas tend to have higher death rates than metropolitan areas, which has been termed "non-metropolitan mortality penalty." Therefore, failure to control for residence in urban areas can also confound associations between air pollution and mortality (Jerrett et al., 2013).

In epidemiological analysis, indicator pollutants are the species selected to represent a mixture of hundreds of gaseous species and particles of complex physico-chemical composition (Brauer et al., 2012). However, inclusion of a single indicator pollutant in the analysis does not provide a realistic estimate, due to the fact that air pollution refers to a complex mixture of gaseous species and particles that permanently interact and undergo transformations. For instance, Janssen et al., (2011) suggest that NO_2 should not be taken as a reliable indicator of traffic due to its high background concentrations and that other species, including BC (black carbon) can better reflect health risk of combustion particles from traffic, solid fuel burning, shipping and industrial processes. They showed that an increase in life expectancy due to hypothetical traffic abatement measures was four to nine times higher when the calculation included BC compared with equivalent decrease in $PM_{2.5}$ concentrations. Furthermore, co-exposure to noise, as the potential confounding factor in traffic pollution studies, has been shown to be related to circulatory disease.

Even if each person's exposure to air pollution could be accurately determined, the pollutant concentrations would not be the only indicator of their potency. Depending on the sampling site, particulate matter exhibits significant differences in chemical composition and related toxicity. For instance, it has been shown that the carbonaceous particles derived from fossil fuel burning are more toxic than crustal material and secondary aerosols (Lelieveld et al., 2015). The presence of thousands of non-volatile and semi-volatile organic compounds in particles also remains an issue insufficiently addressed in exposure science (Solomon et al., 2012). For instance, the recent study of Raaschou-Nielsen et al., (2016) pointed out to the association between sulfur and nickel content in particles and lung cancer incidence. However, since sulfur is not known to be carcinogenic, the authors assume that it might represent a mixture of combustion-related organics or polycyclic aromatic hydrocarbons.

There is scarce evidence regarding the association between air pollution and mortality modified by socio-economic status, education and demographic factors. Krewski et al., (2000) conducted an independent re-analysis project to identify covariates that may confound previously reported relationship between air pollution and mortality. They found a significant modifying effect of education. Some studies also found significantly higher effect estimates among obese participants and those with lower educational attainment and/or socio-economic status (Brunekreef et al., 2009; Ostro et al., 2009). However, Samet et al., (2000) found no evidence that socio-economic factors, such as poverty and race, are significant modifiers of the PM_{10} effect estimates. Similar findings have also been reported by other authors.

The health effects of other environmental stressors, such as temperature, have been widely investigated. Nonetheless, a relatively small number of studies deal with the interacting effects of air pollution and extreme temperature. The study of Meng et al., (2012), including PM_{10} data from eight Chinese cities with average daily concentrations in the range from 65 to 124 $\mu g\ m^{-3}$, showed that a 10 $\mu g\ m^{-3}$ increment in PM_{10} concentrations on high temperature days corresponded to a more than doubled relative risk for air pollution-related mortality, whereas no significant effect modification was observed during cold days. Similarly, the study of Chen et al., (2015) that investigated seasonality of the pollution-attributable death rates in 17 Chinese cities, showed that relative risk associated with a 10-$\mu g\ m^{-3}$ increase in two-day average PM_{10} concentrations was highest in summer and winter months. Vanos et al., (2014) also reported significant modification of air pollution-related mortality risk by weather type and season. Namely, the estimated effect was largest during tropical days in the spring or summer. This can be explained by the temperature-related physiological stress which makes people more susceptible to the adverse effects of pollution.

In conclusion to this, air pollution commonly used in retrospective epidemiological studies is a contextual risk factor, together with temperature, which represents the living environment and explains spatial mortality variations (Sheppard et al., 2012). Nevertheless, the potential confounders are associated with individual factors, e.g., smoking, alcohol use, obesity, presence of different clinical conditions and consumption of anti-hypertensives, lipid-lowering agents, and antidiabetic agents, which also play an important role in multi-causal diseases and the observed death rates.

CONCLUSION

In this study, we discuss the effects of air pollution exposure on death rates in Belgrade, on the basis of the 6-year (2009-2014) regular pollutant monitoring and the corresponding administrative records on mortality. The poor health status of the examined population is reflected in high circulatory and malignant death rates, as well as in the fact that more than one third of malignant deaths are observed among the population younger than 65 years, and that the estimated life expectancy of 74.3 years is considerably lower than the EU average of 80.6 years. In Belgrade area, air pollution represents a significant environmental factor affecting human health. The number of days with average PM_{10} concentrations exceeding 50 $\mu g\ m^{-3}$ during the study period was in the range from 75 to 155 per year, which is considerably higher than the margin of tolerance set out in the Air Quality Standards. According to the results, no substantial association was observed between mortality and short-term exposure: a 10 $\mu g\ m^{-3}$ increase in the average pollutant concentration is estimated to cause a negligible increase in death counts, rarely exceeding 0.5%. Statistically significant associations were detected between PM_{10} and SO_2 and circulatory mortality in elderly and males, and SO_2 and malignant mortality in elderly and female population. However, the results further suggest that the exposure over one-year period preceding death might be more significant than short-term exposure. The elevated concentrations of NO_2 were associated with a significant increase in circulatory and malignant mortality in people older than 65 years, and particularly male population. In general, the effects of PM_{10} were estimated as less significant compared to the effects of gaseous oxides.

The health effects of air pollution follow a complex pattern across different geographic areas, depending on the population features. Even though individual studies are necessary to estimate the magnitude and trend of such relationships and to extend the current knowledge on how diverse populations respond to environmental stressors, they are mostly based on a limited data and their findings can be generalized only upon confirmation through meta-analyses. In addition, such analyses are used to summarize the results of relevant studies and reach the conclusion through a systematic approach.

ACKNOWLEDGMENTS

This paper was realized as part of projects No III43007 and No III41011 founded by the Ministry of Education, Science and Technological Development of the Republic of Serbia for the period 2011-16. The authors are thankful to the Institute of Public Health in Belgrade for providing the necessary facilities and data.

REFERENCES

Allen, RW; Gombojav, E; Barkhasragchaa, B; Byambaa, T; Lkhasuren, O; Amram, O; Takaro, TK; Janes, CR. *Air Qual Atmos Health*, 2013, 6, 137-150.

Anenberg, SC; Horowitz, LW; Tong, DQ; West, *J. Environ Health Perspect.*, 2010, 118, 1189-1195.

Basu, R; Samet, JM. *Epidemiol Rev.*, 2002, 24, 190-202.

Beelen, R; Raaschou-Nielsen, O; Stafoggia, M; Andersen, ZJ; Weinmayr, G; Hoffmann, B; Wolf, K; Samoli, E; Fisher, P; Nieuwenhuijsen, M; Vineis, P; Xun, WW; Katsouyanni, K; Dimakopolou, K; Oudin, A; Forsberg, B; Modig, L; Havulinna, AS; Lanki, T; Turunen Oftedal, B; Nystad, W; Nafstad, P; De Faire, U; Pedersen, NL; Östenson, CG; Fratiglioni, L; Penell, J; Korek, M; Pershagen, G; Eriksen, KT; Overvad, K; Ellermann, T; Eeftens, M; Peeters, P. H; Meliefste, K; Wang, M; Bueno-de-Mesquita, B; Sugiri, D; Krämer, U; Heinrich, J; De Hoogh, K; Key, T; Peters, A; Hampel, R; Concin, H; Nagel, G; Ineichen, A; Schaffner, E; Probst-Hensch, N; Künzli, N; Schindler, C; Schikowski, T; Adam, M; Phuleria, H; Vilier, A; Clavel-Chapelon, F; Declercq, C; Grioni, S; Krogh, V; Tsai, MY; Ricceri, F; Sacerdote, C; Galassi, C; Migliore, E; Ranzi, A; Cesaroni, G; Badaloni, C; Forastiere, F; Tamayo, I; Amiano, P; Dorronsoro, M; Katsoulis, M; Trichopoulou, A; Brunekreef, B; Hoek, G. *Lancet*, 2014, 383, 785-795.

Brauer, M; Amann, M; Burnett, RT; Cohen, A; Dentener, F; Ezzati, M; Henderson, SB; Krzyzanowski, M; Martin, RV; Van Dingenen, R; Van Donkelaar, A. *Environ Sci Technol.*, 2012, 46, 652-660.

Bräuner, EV; Forchhammer, L; Møller, P; Simonsen, J; Glasius, M; Wåhlin, P; Raaschou-Nielsen, O; Loft, S. *Environ Health Perspect.*, 2007, 115, 1177–1182.

Brook, JR; Burnett, RT; Dann, TF; Cakmak, S; Goldberg, MS; Fan, X; Wheeler, AJ. *J Expo Sci Env Epid.*, 2007, 17, S36-S44.

Brook, RD; Rajagopalan, S; Pope, CA; Brook, JR; Bhatnagar, A; Diez-Roux, AV; Holguin, F; Hong, Y; Luepker, RV; Mittleman, MA; Peters, A. *Circulation*, 2010, 121, 2331-2378.

Brunekreef, B; Beelen, RMJ; Hoek, G; Schouten, LJ; Bausch-Goldbohm, S; Fischer, P; Armstrong, B; Hughes, E; Jerrett, M. *Research report (Health Effects Institute)*, 2009, 139, 5-71.

Cancer Research, UK. Cancer Mortality by Age. (2015). http://www.cancerresearchuk.org/health-professional/cancer-statistics/mortality/age#heading-Zero.

Carey, IM; Atkinson, RW; Kent, AJ; Van Staa, T; Cook, DG; Anderson, HR. *Am J Respir Crit Care Med.*, 2013, 187, 1226-1233.

Carslaw, DC; Ropkins, K. *Environ Modell Softw.*, 2012, 27, 52-61.

Chen, H; Goldberg, MS; Burnett, RT; Jerrett, M; Wheeler, AJ; Villeneuve, PJ. *Epidemiology*, 2013, 24, 35-43.

Chen, R; Kan, H; Chen, B; Huang, W; Bai, Z; Song, G; Pan, G. *Am J Epidemiol.*, 2012, doi: 10.1093/aje/kwr425.

Chen, R; Peng, RD; Meng, X; Zhou, Z; Chen, B; Kan, H. *Sci Tot Environ.*, 2013, 450, 259-265.

Chen, Y; Ebenstein, A; Greenstone, M; Li, H. *PNAS*, 2013, 110, 12936-12941.

Clancy, L; Goodman, P; Sinclair, H; Dockery, DW. *Lancet*, 2002, 360, 1210-1214.

Correia, AW; Pope III, CA; Dockery, DW; Wang, Y; Ezzati, M; Dominici, F. *Epidemiology*, 2013, 24, 23-31.

DeMarini, D. M. *Mutagenesis*, 2013, 28, 485-505.

Detailed Mortality Database. World Health Organization. Regional Office for Europe. (2015). http://data.euro.who.int/dmdb/.

Dimakopoulou, K; Samoli, E; Beelen, R; Stafoggia, M; Andersen, ZJ; Hoffmann, B; Fischer, P; Nieuwenhuijsen, M; Vineis, P; Xun, W; Hoek, G. *Am J Respir Crit Care Med.*, 2014, 189, 684-696.

Evans, J; van Donkelaar, A; Martin, RV; Burnett, R; Rainham, DG; Birkett, NJ; Krewski, D. *Environ Res.*, 2013, 120, 33-42.

Fang, Y; Mauzerall, DL; Liu, J; Fiore, AM; Horowitz, LW. *Climatic Change*, 2013, 121, 239-253.

Fischer, H; Wagenbach, D; Kipfstuhl, J. *J Geophys Res Atmos.*, 1998, 103, 21927-21934.

Gasparrini, A. *J Stat Softw.*, 2011, 43, 1-20.

Gasparrini, A; Armstrong, B; Kenward, MG. *Stat Med.*, 2010, 29, 2224-2234.

Gasparrini, A; Guo, Y; Hashizume, M; Lavigne, E; Zanobetti, A; Schwartz, J; Tobias, A; Tong, S; Rocklöv, J; Forsberg, B; Leone, M. *Lancet*, 2015, 386, 369-375.

Giannadaki, D; Pozzer, A; Lelieveld, J. *Atmos Chem Phys.*, 2014, 14, 957-968.

Goodman, PG; Dockery, DW; Clancy, L. *Environ Health Perspect.*, 2004, 112, 179-185.

Heinrich, J; Hoelscher, B; Frye, C; Meyer, I; Pitz, M; Cyrys, J; Wjst, M; Neas, L; Wichmann, HE. *Epidemiology*, 2002, 13, 394-401.

Hoek, G; Krishnan, RM; Beelen, R; Peters, A; Ostro, B; Brunekreef, B; Kaufman, J. D. *Environ Health*, 2013, 12, 43-58.

Janssen, NA; Hoek, G; Simic-Lawson, M; Fischer, P; Van Bree, L; Ten Brink, H; Keuken, M; Atkinson, RW; Anderson, HR; Brunekreef, B; Cassee, FR. *Environ Health Perspect.*, 2011, 119, 1691-1699.

Jerrett, M; Burnett, RT; Beckerman, BS; Turner, MC; Krewski, D; Thurston, G; Martin, RV; van Donkelaar, A; Hughes, E; Shi, Y; Gapstur, SM. *Am J Respir Crit Care Med.*, 2013, 188, 593-599.

Krewski, D; Burnett, RT; Goldberg, MS; Hoover, K; Siemiatycki, J; Jerrett, M; Abrahamowicz, M; White, WH. Reanalysis of the Harvard Six Cities Study and the American Cancer Society Study of particulate air pollution and mortality; Health Effects Institute: Cambridge, 2000.

Krewski, D; Jerrett, M; Burnett, RT; Ma, R; Hughes, E; Shi, Y; Turner, M. C; Pope III, CA; Thurston, G; Calle, EE; Thun, MJ. Extended follow-up and spatial analysis of the American Cancer Society study linking particulate air pollution and mortality (No. 140); Health Effects Institute: Boston, 2009.

Lelieveld, J; Evans, JS; Fnais, M; Giannadaki, D; Pozzer, A. *Nature*, 2015, 525, 367-371.

Lepeule, J; Laden, F; Dockery, D; Schwartz, J. *Environ Health Perspect.*, 2012, 120, 965-970.

Medley, AJ; Wong, CM; Thach, TQ; Ma, S; Lam, TH; Anderson, HR. *Lancet*, 2002, 360, 1646-1652.

Meng, X; Zhang, Y; Zhao, Z; Duan, X; Xu, X; Kan, H. *Sci Tot Environ.*, 2012, 435, 215-221.

Milojević, A; Wilkinson, P; Armstrong, B; Bhaskaran, K; Smeeth, L; Hajat, S. *Heart*, 2014, 100, 1093-1098.

Mortality and life expectancy statistics. (2013). Eurostat, Statistics Explained. http://ec.europa.eu/eurostat/statistics-explained/index.php/Mortality_and _ life_expectancy_statistics.

Noonan, CW; Ward, TJ; Navidi, W; Sheppard, L. *Occup Environ Med.*, 2012, 69, 354-60.

Ostro, BD; Feng, WY; Broadwin, R; Malig, BJ; Green, RS; Lipsett, MJ. *Occup Environ Med.*, 2008, 65, 750-756.

Parker, JD; Mendola, P; Woodruff, TJ. *Epidemiology*, 2008, 19, 820-823.

Perišić, M; Stojić, A; Stojić, SS; Šoštarić, A; Mijić, Z; Rajšić, S. *Air Qual Atmos Health*, 2014, 1-11.

Peters A. *Toxicol Appl Pharmacol.*, 2005, 207, 477–482.

Pope, CA; Burnett, RT; Thurston, GD; Thun, MJ; Calle, EE; Krewski, D; Godleski, JJ. *Circulation*, 2004, 109, 71-77.

Pretty, R. (2015). TheilSen {openair} Tests for trends using Theil-Sen estimates. http://www.inside-r.org/packages/cran/openair/docs/TheilSen.

R Core Team. (2014). R: A Language and Environment for Statistical Computing. Vienna, Austria: R Foundation for Statistical Computing. http://www.r-project.org.

Raaschou-Nielsen, O; Andersen, ZJ; Jensen, SS; Ketzel, M; Sorensen, M; Hansen, J; Loft, S; Tjonneland, A; Overvad, K. *Environ Health.*, 2012, 11, 860-865.

Raaschou-Nielsen, O; Beelen, R; Wang, M; Hoek, G; Andersen, ZJ; Hoffmann, B; Stafoggia, M; Samoli, E; Weinmayr, G; Dimakopoulou, K; Nieuwenhuijsen, M. *Environ Int.*, 2016, 87, 66-73.

Ross, CE; Masters, RK; Hummer, RA. *Demography*, 2012, 49, 1157-1183.

Salam, A; Albinali, H; Singh, R; Asaad, N; Alqahtani, A; Al Suwaidi, J. *J Am Coll Cardiol.*, 2013, 168, 975–980.

Samet, JM; Zeger, SL; Dominici, F; Curriero, F; Coursac, I; Dockery, DW; Schwartz, J; Zanobetti, A. The national morbidity, mortality, and air pollution study. Part II: morbidity and mortality from air pollution in the United States; Health Effects Institute: Cambridge, 2000.

Schindler, C; Keidel, D; Gerbase, MW; Zemp, E; Bettschart, R; Brandli, O; Brutsche, MH; Burdet, L; Karrer, W; Knopfli, B; Pons, M. *Am J Respir Crit Care Med.*, 2009, 179, 579-587.

Seaton, A; Godden, D; MacNee, W; Donaldson, K. *Lancet*, 1995, 345, 176–178.

Shah, AS; Langrish, JP; Nair, H; McAllister, DA; Hunter, AL; Donaldson, K; Newby, DE; Mills, NL. *Lancet*, 2013, 382, 1039-1048.

Shang, Y; Sun, Z; Cao, J; Wang, X; Zhong, L; Bi, X; Li, H; Liu, W; Zhu, T; Huang, W. *Environ Int.*, 2013, 54, 100-111.

Shaposhnikov, D; Revich, B; Bellander, T; Bedada, GB; Bottai, M; Kharkova, T; Kvasha, E; Lezina, E; Lind, T; Semutnikova, E; Pershagen, G. *Epidemiology*, 2014, 25, 359.

Sheppard, L; Burnett, RT; Szpiro, AA; Kim, SY; Jerrett, M; Pope III, CA; Brunekreef, B. *Air Qual Atmos Health*, 2012, 5, 203-216.

Silva, RA; West, JJ; Zhang, Y; Anenberg, SC; Lamarque, JF; Shindell, DT; Collins, WJ; Dalsoren, S; Faluvegi, G; Folberth, G; Horowitz, LW. *Environ Res Lett.*, 2013, 8, 034005.

Solomon, PA; Costantini, M; Grahame, TJ; Gerlofs-Nijland, ME; Cassee, FR; Russell, AG; Brook, JR; Hopke, PK; Hidy, G; Phalen, RF; Saldiva, P; Sarnat, SE; Balmes, JR; Tager, IB; Özkaynak, H; Vedal, S; Wierman, SSG; Costa, DL. *Air Qual Atmos Health*, 2012, 5, 9-62.

Šoštarić, A; Mladenović, S; Stanišić Stojić, S; Stojić, A; Stanišić, N; Slepčević V. Health burden of air pollutant exposure in Belgrade: a European region with high circulatory and malignant mortality rates, Newsletters, WHO Collaborating Centre for Air Quality Management and Air Pollution Control, No. 56 - December 2015, 3-9. http://envpl.ipb.ac.rs/pages/newsletters/WHOdecember_2015_nl56_1.pdf.

Stanišić Stojić, S; Stanišić, N; Stojić, A; Šoštarić, A. *J Toxicol Environ Health A*, 2016, 79, 17-27.

Statistical Office of the Republic of Serbia. 2015. The additionally processed mortality data. Vital statistics.

Stojić, A; Stanišić Stojić, S; Mijić, Z., Ilić, L; Tomašević, M; Todorović, M; Perišić, M. In Comprehensive Analysis of VOC Emission Sources in Belgrade Urban Area, In: Urban and Built Environments: Sustainable Development, Health Implications and Challenges; Cohen, A.; Ed.; Nova Science Publishers: New York, USA, 2015c; pp. 55-88.

Stojić, A; Stanišić Stojić, S; Reljin, I; Čabarkapa, M; Šoštarić, A; Perišić, M; Mijić, Z. *Environ Sci Poll Res.*, 2016, 1-11. doi: 10.1007/s11356-016-6266-4.

Stojić, A; Stojić, SS; Mijić, Z; Šoštarić, A; Rajšić, S. *Atmos Environ.*, 2015a, 106, 71-79.

Stojić, A; Stojić, SS; Šoštarić, A; Ilić, L; Mijić, Z; Rajšić, S. *Environ Sci Poll Res.*, 2015b, 22(17), 13137-13152.

Tanaka, S. *J Health Econ.*, 2015, 42, 90-103.

Todorović, M; Perišić, M; Kuzmanoski, M; Stojić, A; Šoštarić, A; Mijić, Z; Rajšić, S. *J Environ Sci Heal A.*, 2015, 50(13), 1351-1359.

Turner, MC; Krewski, D; Pope III, CA; Chen, Y; Gapstur, SM; Thun, MJ. *Am J Respir Crit Care Med.*, 2011, 184, 1374-1381.

Vanos, JK; Hebbern, C; Cakmak, S. *Environ Pollut.*, 2014, 185, 322-332.

Ye, X; Wolff, R; Yu, W; Vaneckova, P; Pan, X; Tong, S. *Environ Health Perspect.*, 2012, 120, 19-28.

Zhang, Z; Fang, J; Gillespie, C; Wang, G; Hong, Y; Yoon, PW. *Am J Cardiol.*, 2012, 109, 1097-1103.

In: Air Pollution
Editor: Gerald L. Burns
ISBN: 978-1-63485-374-3

Chapter 4

BIOAEROSOLS: METHODS FOR REDUCING HEALTH RISKS AND IMPACT ON THE ENVIRONMENT

Snežana M. Šerbula[1,*], Jelena S. Milosavljević[1], Tanja S. Kalinović[1], Ana A. Radojević[1], Jelena V. Kalinović[1], Branko M. Bugarski[2] and Jasmina S. Stevanović[3]

[1]University of Belgrade, Technical Faculty in Bor, Bor, Serbia
[2]University of Belgrade, Faculty of Technology and Metallurgy, Belgrade, Serbia
[3]Institute for Chemistry, Technology and Metallurgy, Belgrade, Serbia

ABSTRACT

Bioaerosols (biological particles derived from viruses, bacteria, fungi, plants and animals) are ubiquitous in the air. This organic dust is considered to be hazardous to human health. Acute and chronic diseases are health risks associated with viable and non-viable airborne biological particles because they can be toxic, allergenic and infectious. With increasing awareness of the environmental impact of bioaerosols, which can spread easily with air currents over long distances, the development of efficient air filtration system is important for preventing health risks.

[*] Corresponding author: E–mail address: ssherbula@tf.bor.ac.rs.

Air filtration system is the most used method for controlling air pollution with bioaerosols but it is not always efficient because microbes can survive and proliferate on the surface of filters and be potential sources of diseases and allergies. Lately, many researchers have focused on improving indoor air quality by removing and inactivating biological agents. Successful methods for control and inactivation of bioaerosols include filters coated with antimicrobial nanoparticles from different plants, such as *Sophora flavescens, Euscapis japonica* or *Melaleuca alternifolia.* Filters coated with plant extracts have filtration efficiency from 77 to 99%, while the range for antimicrobial activity for these filters is from 14 to 100%. These natural products are considered ecological, environmentally friendly, cost–effective, less toxic (compared with inorganic nanomaterials) and efficient material for antimicrobial filter media development and can be easily applied to conventional air–conditioning systems.

1. INTRODUCTION

Bioaerosols or organic dust (Douwes et al., 2003) are defined as airborne particles with biological origin, which include bacteria, viruses, fungi, fungal spores, pollen and pollen fragments, numerous antigenes (Małecka–Adamowicz et al., 2011), yeast, protozoans (Tamer Vestlund et al., 2014) and algae (Van Leuken et al., 2016). Arthropods, microbial products (β–1,3–glucans, endotoxins, enzymes) (Taha et al., 2005), volatile organic compounds (VOCs) (Sykes et al., 2011), plant and animal fragments (Aríngoli et al., 2008; Šerbula et al., 2013a) are also considered as bioaerosols (Sykes et al., 2011; Šerbula et al., 2014a). 25% of aerosolized matter has biological origin (Woo et al., 2013; Šerbula et al., 2015).

Plants, animals, water and soil are sources of bioaerosols (Heo et al., 2014; Kalinović et al., 2016). It is estimated that concentration of bacteria and fungi in soil are 10^{10}, 10^5 to 10^6 cell/g, respectively (Heikkinen et al., 2005). Organic and inorganic aerosols are present in the air due to different anthropogenic activities such as agricultural production (Šerbula et al., 2012a; Hurtado et al., 2014) and fermentation processes (Wang et al., 2010). Formation and diffusion of pathogenic and non–pathogenic microbes in the air is a consequence of many industrial processes (Grisoli et al., 2009; Šerbula et al., 2011), in particular the processes of wastewater treatment (Orsini et al., 2002; Šerbula et al., 2013b). Daily and seasonal variation of airborne organic and inorganic matter is influenced by many environmental factors (Abdel Haamed et al., 2009; Šerbula et al., 2013c).

Although the method for removing aerosolized matter from the indoor air is economical and efficient, air filters cannot inactivate the captured bioaerosols. Microbes remain viable on filters and represent potential health risk. Applying natural products as antimicrobial agents is less toxic than inorganic materials used for microbe inactivation (Espirirtu et al., 2014).

2. Microbes in the Air

The atmosphere is an environment in which the diversity of microbes varies in space and time, like in all microbial habitats (Yadav et al., 2015). The transport, horizontal and vertical (Šerbula et al., 2014b; Bryan et al., 2014), of airborne biological particles over large distances depends upon environmental conditions (temperature, humidity, wind direction and velocity) and physical factors (tall buildings and trees) (Stellacci et al., 2010). Bioaerosols can spend seconds or even weeks in the air, while their transport can vary from a couple of meters to thousands of kilometers (Sialve et al., 2015). Bacteria and fungi have been found in different atmospheric layers such as the boundary layer at altitude of 1.5 km, the upper troposphere at 12 km altitude and the stratosphere (altitude of 20 and 41 km). *Penicillium notatum* have been isolated at an altitude of 77 km, while *Microccocus albus* and *Mycobacterium luteum* have been found at an altitude of 70 km (Polymenakou, 2012).

The atmosphere is a very stressful environment for microorganisms (Polymenakou, 2012). Besides the availability of sources of nutrients (Aríngoli et al., 2008), the concentration of airborne microbes depends upon vegetation, meteorological, climatic conditions (Šerbula et al., 2012b; Haas et al., 2014) and ultraviolet radiation (Niazi et al., 2015). One of the limiting factors for survival of cells in the air is low temperature. Microbes can adopt different survival mechanisms because their genomes contain the direction for many biochemical transformations. Besides temperature, factors that can limit the survival of aerosolized microorganisms are the presence of oxidants and solar radiation. The expression of different enzymes enables them to reduce radicals, while the production of pigments which absorb UV–B to red wavelengths helps them to avoid damage on the DNA (Polymenakou, 2012). Pollution of the air (carbon monoxide, hydrocarbons, NO_2 and SO_2) can affect microbial survival. The concentration of fungal aerosols decreases with increasing SO_2 (Šerbula et al., 2010; Abdel Hameed et al., 2012).

Properties of bioaerosol such as shape, surface characteristics, affinity for aggregation have the influence on their behavior and can be used as factors for dispersal prediction (Antonijević et al., 2008; Tamer Vestlund et al., 2014). Another parameter which defines the dynamics of bioaerosol behaviour is the size (Alagić et al., 2013; Galès et al., 2015) which can vary from 1 nm to 100 μm (Nasir et al., 2012). Anemophilous plants have pollen with diameter of 17–58 μm, fungal spores have diameter of 1–30 μm, while diameters of bacteria and viruses are 0.25–8 μm and 0.3 μm, respectively (Jones and Harrison, 2004). Particles of biological origin with the size from 1.0 to 5.0 μm usually remain in the air, while particles with larger size shortly get deposited on surfaces (Ghosh et al., 2015). Airborne microbes often form aggregates or can be attached to non–viable particles (Henningson and Ahlberg, 1994).

Atmospheric and indoor conditions can influence the prevalence and distributions of bioaerosols. The amount of charge on particles depends upon environmental conditions and the generation mechanism and can enhance their deposition on oppositely charged surfaces (Heikkinen et al., 2005). Besides the impact on the ecosystem and public health, organic dust can have influence on meteorology and global climate (Polymenakou, 2012).

3. Wastewater Treatment Plants as Sources of Airborne Microorganisms

Physical, chemical and biological treatment is performed in wastewater treatment plants (WWTPs) for reduction and elimination of contaminants before releasing water into the environment (Masclaux et al., 2013, Šerbula et al., 2016). Particularly, aeration and mechanical agitation represent the important sources for aerosolization of microbes (Sánchez–Monedero et al., 2008). The releasing of particles from a surface happens when external forces become stronger than bounding forces, and bursting of bubbles is the major aerosol generator (Sialve et al., 2015). Since bioaerosols consist of microbes which can affect health of plant workers and have negatively impact the environment, the increase in research of their emission has been recorded recently (Han et al., 2012).

A certain condition in the air can inhibit the development of microbes originated from sewage. The desiccation, extreme temperatures and solar radiation can cause dying of airborne microbes. Despite this, some microbes have specific mechanisms which help them survive harsh environmental conditions (Korzeniewska et al., 2009).

Due to the richness in organic matter, wastewaters and sewage sludge have different micro and macroflora representatives (Kacprzak et al., 2005), like *Enterobacteriaceae* bacteria which are naturally present in human gastrointestinal tract (Korzeniewska and Harnisz, 2012). The suggested threshold limits for gram–negative bacteria, total bacteria and total fungi are 300, 1000 and 1000 colony forming units per cubic meter (CFU/m^3), respectively (Drew et al., 2009), while for *Aspergillus fumigatus* is 500 CFU/m^3 (Pankhurst et al., 2011b).

Microorganism counts in the air is one of the indicators of atmospheric pollution and for that reason researches were performed in many WWTP as well as in plants for waste treatment. Concentration of different bioaerosols was measured in WWTPs and in waste treatment plant, such as heterotrophic bacteria (Małecka–Adamowicz et al., 2011; Pascual et al., 2003; Filipkowska et al., 2000, Filipkowska et al., 2002; Gotkowska–Płachta et al., 2013), Staphylococci (Małecka–Adamowicz et al., 2011; Michałkiewicz et al., 2011; Kruczalak and Olańczuk–Neyman, 2004; Filipkowska et al., 2000, Filipkowska et al., 2002; Gotkowska–Płachta et al., 2013), *Pseudomonas fluorescens* (Małecka–Adamowicz et al., 2011; Michałkiewicz et al., 2011; Kruczalak and Olańczuk–Neyman, 2004; Filipkowska et al., 2002), *P. aeruginosa* (Pascual et al., 2003), Pseudomonas genus (Breza–Boruta and Paluszak, 2007; Filipkowska et al., 2000), hemolitic bacteria (Filipkowska et al., 2000, Filipkowska et al., 2002), total coliform (Michałkiewicz et al., 2011; Kruczalak and Olańczuk–Neyman, 2004; Pascual et al., 2003; Karra and Katsivela., 2007; Haas et al., 2010; Filipkowska et al., 2002; Fracchia et al., 2006b), fecal coliform (Pascual et al., 2003; Heinonen–Tanaski et al., 2009; Karra and Katsivela, 2007; Vítězová et al., 2012), enterococci (Karra and Katsivela, 2007; Heinonen–Tanaski et al., 2009; Gotkowska–Płachta et al., 2013; Filipkowska et al., 2002), *Enterococcus* spp. (Breza–Boruta and Paluszak, 2007), *Salmonella* spp. (Breza–Boruta and Paluszak, 2007), *Enterobacteriaceae* (Korzeniewska et al., 2009; Filipkowska et al., 2000; Gotkowska–Płachta et al., 2013), *Escherichia coli* (Breza–Boruta and Paluszak, 2007; Filipkowska et al., 2002; Fracchia et al., 2006b), sulfite–reducers (Fracchia et al., 2006b), *Aspergillus fumigatus* (Sánchez–Monedero and Stentiford, 2003; Taha et al., 2007, Taha et al., 2006; Vilavert et al., 2009; Pankhurst et al., 2011a; Persoons et al., 2010; Abdel Hameed et al., 2015; Haas et al., 2010; Sánchez–Monedero et al., 2005), *Aspergillus* sp. (Persoons et al., 2010, Huang et al., 2002), *Penicillium* sp. (Huang et al., 2002), *Cladosporium* sp. (Huang et al., 2002), actinomycetes (Małecka–Adamowicz et al., 2011; Li et al., 2013b; Breza–Boruta and Paluszak, 2007; Korzeniewska et al., 2009; Prażmo et al., 2003; Taha et al., 2007, Taha et al., 2006; Lehtinen et al., 2013; Persoons et al., 2010; Vítězová et al., 2012; Fracchia et al., 2006a, 2006b; Uhrbrand et al., 2011; Filipkowska et al., 2000, Filipkowska et al., 2002; Gotkowska–Płachta et al., 2013; Li et al., 2012), streptomycetes (Fracchia et al., 2006a, 2006b), fungi spores (Kruczalak and Olańczuk–Neyman, 2004), molds and yeasts (Pascual et al., 2003), moulds (Małecka–Adamowicz et al., 2011; Uhrbrand et al., 2011; Korzeniewska et al., 2009; Filipkowska et al., 2000, 2002; Gotkowska–Płachta et al., 2013), yeasts (Huang et al., 2002; Filipkowska et al., 2000, 2002; Gotkowska–Płachta et al., 2013; Korzeniewska et al., 2009), Gram positive (Solans et al., 2007) and Gram negative bacteria (Prażmo et al., 2003; Vilavert et al., 2009; Oppliger et al., 2005; Persoons et al., 2010; Solans et al., 2007; Filipkowska et al., 2002; Fracchia et al., 2006b), yeasts and filamentous fungi (Korzeniewska et al., 2009).

Table 1 represents concentrations (CFU/m^3) of different airborne microorganisms emitted from wastewater and waste treatment plants.

Table 1. Concentration (Colony Forming Units) of different microorganisms in m^3 of the air

Microorganism	Sampling sites	Concentration (CFU/m^3)		Reference
		Range	Mean value	
Fungi	1	0–265,000***	-	Michałkiewicz et al., (2011)
	2	160–65,700***	-	
	3	280–46,600***	-	
	4	200–44,100***	-	
	5	822–2,156	-	Li et al., (2013b)
	6	113*–21,800***	-	Breza–Boruta and Paluszak (2007)
	7	1,400–2,400$^{\#}$ <100–113$^{\#\#\#}$	1700$^{\#}$ 45$^{\#\#\#}$	Bauer et al., (2002)
	8	10–150	58	Prażmo et al., (2003)
	9	382–3,882 (at 25°C) 18–790 (at 37°C)	-	Vilavert et al., (2009)
	10	470–290,000$^{\#}$	-	Lehtinen et al., (2013)
	11	25–32,000	-	Vítězová et al., (2012)
	12	25–7,950^{i} 0–4,825^{o}	-	Oppliger et al., (2005)
	14	1,000–25,000$^{\#}$	-	Rahkonen (1992)

Microorganism	Sampling sites	Concentration (CFU/m^3)		Reference
		Range	Mean value	
	15	100–10,000^{m}	-	Persoons et al., (2010)
	16	141–1,590	-	Li et al., (2011)
	17	169–833^{i}	-	Ding et al., (2015)
	18	2,584–11,746	-	Huang et al., (2002)
	19	853–4,280$^{\#}$ 0–3,910$^{\#\#}$	1954$^{\#}$ 1,332.6$^{\#\#}$	Abdel Hameed et al., (2015)
	20	1,060–>12,000	-	Solans et al., (2007)
	21	4,800–103,200	-	Lavoie et al., (2006)
	22	829–1,200^{d} 802–1,198^{e} 868–1,289^{f} 279–516^{g}	-	Maharia and Srivastava (2015)
	23	20–930	-	Li et al., (2015)
Bacteria	1	0*–195,000$^{***;\#}$ 78–225,000^{p}	-	Michałkiewicz et al., (2011)
	2	160*–34,000$^{***;\#}$ 200–84,200^{P}	-	
	3	0*–5,350$^{***;\#}$ 650–15,100^{P}	-	
	4	52*–8,350$^{***;\#}$ 300–29,300^{P}	-	
	24	1–1,324$^{IM;\#}$ several tens–9,430$^{SM;\#}$ 14–5,255$^{IM;p}$ 157–18,250$^{SM;p}$	-	Kruczalak and Olańczuk–Neyman (2004)
	5	1,696–7,866	-	Li et al., (2013b)
	6	41*–27,100***	-	Breza–Boruta and Paluszak (2007)
	7	13,000–24,000$^{\#}$	17,000$^{\#}$	Bauer et al., (2002)
	8	240–7,070$^{\#}$	1,955$^{\#}$	Prażmo et al., (2003)
	25	2,300–110,000$^{\#}$	-	Sánchez–Monedero and Stentiford (2003)
	9	44–926 (at 37°C)	-	Vilavert et al., (2009)
	10	480–220,000$^{\#}$	-	Lehtinen et al., (2013)
	11	20–18,500$^{\#}$ 14–12,000^{p}	-	Vítězová et al., (2012)
	26	261–4,350$^{\#}$ (at 22°C) ND–5,360$^{\#}$ (at 37°C) 8–>40,000$^{\#\#}$	-	Fracchia et al., (2006a)

Table 1. (Continued)

Microorganism	Sampling sites	Concentration (CFU/m^3)		Reference
		Range	Mean value	
	27	189–>40,000$^{22°C;\#}$ 200–>40,000$^{37°C;\#}$ 28–>40,000$^{48°C;\#\#}$	-	
	12	2,225–26,200^{i} 75–8,775^{o}	-	Oppliger et al., (2005)
	14	600–14,000$^{\#}$ 20–450$^{\#\#\#}$	-	Rahkonen (1992)
	15	100–10,000	-	Persoons et al., (2010)
	16	459–5,565	-	Li et al., (2011)
	17	1,458–4,155^{i}	-	Ding et al., (2015)
	28	140–4,580	-	Sánchez–Monedero et al., (2008)
	18	1,160–6,495	-	Huang et al., (2002)
	29	33–>40,000$^{\#}$ (at 22°C) 167–3,922$^{\#}$ (at 37°C) 239–>40,000$^{\#\#}$ (at 48°C)	-	Fracchia et al., (2006b)
	30	72–5,461$^{\#}$ (at 22°C) 39–18,700$^{\#}$ (at 37°C) 61–3,122 $^{\#\#}$ (at 48°C)	-	
	31	261–1,994$^{\#}$ (at 22°C) 300–3,433$^{\#}$ (at 37°C) 372–6,278$^{\#\#}$ (at 48°C)	-	
	32	ND–1,454,545.45	-	Pankhurst et al., (2011a)
	33	1,800–137,200	-	Heinonen–Tanaski et al., (2009)
	34	ND–3,952	-	Li et al., (2013c)
	35	-	4168$^{a;\#}$ 2422$^{b;\#}$ 1929$^{c;\#}$	Li et al., (2013a)
	36	133–4,044$^{\#}$	-	Uhrbrand et al., (2011)
	37	48–11,000$^{i;\#}$ 4–2,000$^{o;\#}$	-	Haas et al., (2010)
	38	<100–>261,000$^{\#}$	-	Sánchez–Monedero et al., (2005)

Microorganism	Sampling sites	Concentration (CFU/m^3)		Reference
		Range	Mean value	
	21	4,800–89,200	-	Lavoie et al., (2006)
	39	86–7,143	-	Brennen et al., (1988)
	23	233–1,697	-	Li et al., (2015)
	13	1,120–8,942	-	Li et al., (2012)

Sampling sites: 1 – the complex wastewater treatment and wastes composting plant (Grodzisk Mazowiecki, Poland); 2 – WWTP (Słupca, Poland); 3 – WWTP (Września, Poland); 4 – WWTP (Kostrzyń, Poland); 5 – municipal sewage treatment plant (Xi'an, China); 6 – municipal sewage treatment plant (Toruń, Poland); 7 – activated sludge pilot plant (Vienna, Austria); 8 – sewage treatment plant (Poland); 9 – municipal solid waste incinerator (MSWI) (Catalonia, Spain); 10 – solid waste management plants (Hämeenlinna and Hyvinkää, Finland); 11 – WWTP (Czech Republic); 12 – sewage treatment plants (Switzerland); 13 – municipal wastewater treatment plant (Xi'an, China); 14 – waste treatment plants (Finland); 15 – green waste composting unit; 16 – municipal wastewater treatment plant (Beijing, China); 17 – wastewater treatment station (Beijing, China); 18 – municipal landfill – undergoing–closure area (Taiwan); 19 – wastewater treatment plant (Egypt); 20 – household waste plant; 21 – waste collectors (Quebec, Canada); 22 – sewage treatment plant (New Delhi, India); 23 – wastewater treatment plant (Beijing, China); 24 – mechanical–biological wastewater treatment plant (Deborgoze and Wschod, Poland); 25 – organic waste composting plant; 26 – wastewater treatment plant A (Italy); 27 – wastewater treatment plant B (Italy); 28 –wastewater treatment plants (Spain); 29 – composting plant 1 (Italy); 30 – composting plant 2 (Italy); 31 – composting plant 3 (Italy); 32 – composting facility (London, England); 33 – seven wastewater treatment plants (Finland); 34 – municipal wastewater treatment plant (Beijing, China); 35 – wastewater treatment plant (Xi'an, China); 36 – wastewater treatment plant (Copenhagen, Denmark); 37 – sewage system (Austria); 38 – green waste composting plant (England); 39 – wastewater management system (Michigan);

Abbreviations: * – uncontaminated air; ** – moderately contaminated air; ***– strongly contaminated air; IM – impaction method; ^{SM} – sedimentation method; ND – not detected; $^{\#}$ – mesophilic; $^{\#\#}$ – termophilic; $^{\#\#\#}$ – termotolerant; ND – not detected; ^{m} – microscopic; ^{p} – psychrophilic bacteria; ^{i} – indoor; ^{o} – outdoor; $^{a, b, c}$ – 2, 5 and 10 m downwind of the rotating brush, respectively; ^{d} – pre–monsoon season; ^{e} – monsoon; ^{f} – post–monsoon season; ^{g} – winter season; "-" – not available data

4. METHODS FOR REMOVAL AND INACTIVATION OF BIOAEROSOLS

The main factor which determines the efficiency of a specific method for inactivation of microorganisms is microbial response to a specific stress. The research of thermal inactivation of *Bacillus subtilis* var. n*iger* endospores has shown that 99.999% spores was inactivated at 375–400°C due to DNA damage and denaturation of essential proteins (Grinshpun et al., 2010). 99.9% *Escherichia coli* and *Bacillus subtilis* vegetative cells were inactivated at 160°C and 350°C, respectively (Jung et al., 2009c). Inactivation of more than 99% of *Cladosporium cladosporoides* bioaerosols was achieved at 400°C, while 99% of *Aspergillus versicolor* lost its culturability at 350°C (Jung et al., 2009b), and 400°C (Lee et al., 2010b). The main downside of this method is consumption of a relatively large amount of power for maintaining high temperature to inactivate bioaerosols (Hwang et al., 2010). Besides, Johansson et al., (2011) have shown significant association between the inactivation factor and the increase in mutational frequency of *B. atrophaeus* spores DNA. Simultaneous exposition to UV irradiation and thermal energy contributed to inactivation of more than 99% and 98.5% of *Staphyloccocus epidermidis* and *B. subtilis*, respectively. Although UV irradiation requires low energy consumption, this process produces ozone which is associated with different health risks (Hwang et al., 2010).

Ultraviolet germicidal air irradiation (UVGI) is one of technologies that is used for inactivation of microbes. Lamps which are used for UVGI emit nonionizing electromagnetic radiation that may cause DNA damage and interruption of cell replication (Kujundzic et al., 2007). Experiments were performed in order to investigate efficiency of UVGI (Ko et al., 2002; Green and Scarpino, 2002; Kujundzic et al., 2006). For inactivation of 99% of fungi spores than for bacterial cells 80 times bigger dose is necessary (Lin and Li, 2002). Research of Xu et al., (2003) has showed that reduction in bioaerosol concentration by UVGI was: 46–80% for *B. subtilis* spores, 83–98% for *Mycobacterium parafortuitum* and 96–97% for *Mycobacterium bovis* BCG cell. While reductions of 50% for *B. subtilis* and *Microccocus luteus* and 100% for *E. coli* were achieved in a study of Miller and Macher (2000).

Microwave inactivation of biological agents in the air is a simple method for bioaerosol inactivation (Wu and Yao, 2011). Microwave energy has the frequency range from 300 Hz to 300 GHz (Wu and Yao, 2010). Wu and Yao (2010) have showed that inactivation mechanisms for different bacteria vary due to differences in their cell wall structures. Microwave irradiation can be successfully applied in combination with HVAC (heating, ventilation and air conditioning) filter for inactivation of viral aerosols (Woo et al., 2012). Microwave assisted nanofibrous filtration was demonstrated to be effective for disinfection of air contaminated with *E. coli* and *B. subtilis* (Zhang et al., 2010). Microwave inactivation can induce genetic mutations in bacteria which is a downside of this method (Wu and Yao, 2011).

The usage of UV irradiation in combination with a semi–conductor is involved in the photocatalytic process in which free radicals are formed. Titanium dioxide (TiO_2) is the most frequently used semi–conductor and it is present in the form of a photocatalytic nano–powder (Faure et al., 2011). Free radicals formed in photocatalytic oxidation (PCO) attack cell membranes of microbes and induce releasing of K^+, ribonucleic acid (RNA) and proteins, which leads to cell death (Chuaybamroong et al., 2011). The efficiency of PCO depends upon species of bacteriophages (Marcó et al., 2011). The combination of PCO and negative air ionization (NAI) was found to be a good method for removal of aerosolized *E. coli*, *Candida famata* and λ vir phage (Yu et al., 2008). The inactivation of 90% of initially viable MS2 viruses and *B. subtilis* spores was achieved by unipolar ion emission and PCO (Grinshpun et al., 2007). The contact surface between bioaerosols and active photocatalytic coating is an essential factor which affects the success of this process (Keller et al., 2005). Filters which inactivate bioaerosols via photocatalysis are expensive, complex and may produce toxic by–product like phosgene during the process (Pyankov et al., 2008).

Antimicrobial activity of silver is widely known (Sharma et al., 2009). Mechanisms that explain toxicity of silver are: (1) Adenosine triphosphate (ATP) production and deoxyribonucleid acid (DNA) replication are disrupted due to the uptake of silver ions; (2) ROS (Reactive Oxygen Species) generation is induced by silver nanoparticles and ions; (3) direct damage of cell membranes is caused by silver nanoparticles (Marambio–Jones and Hoek, 2010). Airborne silver nanoparticles show potential for controlling bioaerosols (*S. epidermidis*, *E. coli* and *B. subtilis*) deposited on the filter (Lee et al., 2008). Silver nanoparticles can be used with SiO_2 particles for viral inactivation (Joe et al., 2014). Silver can serve as a doping agent for enhancement of photocatalytic activity of TiO_2 coated on glass fibers and inhibition of *B. subtilis* spores (Zacarías et al., 2015), *E. coli* (Pham and Lee, 2014a) and *Staphyloccocus* (Pham and Lee, 2014b). Downsides of the glass fiber as a substrate for the deposition of TiO_2 are low surface area and low adsorption ability so that photocatalytic activity is not maximally exhibited (Pham and Lee, 2015). Excellent adhesion of TiO_2 coatings is necessary in order to prevent loss of a catalyst during the process and avoid health risks and negative environmental impact (Yao and Yeung, 2011). Potentially toxic by–products generated during bioaerosols inactivation in photocatalytic filters need to be identified (Pigeot–Remy et al., 2013). Health risks associated with human exposure to high concentrations of TiO_2 and silver nanoparticles are neurological and respiratory problems (Hwang et al., 2010). Silver nanoparticles can interact with cellular enzymes and proteins involved in cell's antioxidative defense mechanisms which lead to ROS accumulation. Over accumulation of these reactive species can initiate inflammatory response, destruction of mitochondria and finally cell death (Chen and Schluesener, 2008).

Although high efficiency particulate air (HEPA) filter can efficiently capture bioaerosols, they can proliferate on filter. In order to eliminate this problem, the investigation of a combination of PCO and HEPA filter was performed. HEPA filters loaded with TiO_2 were able to inactivate 60–100% of retained microbes (Chuaybamroong et al., 2010). The performance of HVAC (heating, ventilation and air conditioning) filter against viable biological particles is enhanced with continuous emission of unipolar (negative) ions (Huang et al., 2008). Carbon fiber ionizer can reduce survival of *E. coli* (Park et al., 2009), and *E. coli* and *S. epidermidis* (Kim et al., 2011), captured on filters. Lee et al., (2014) have shown that carbon fiber ionizer can inactivate *S. epidermidis* in HVAC system.

4.1. Filters Coated with Natural Antimicrobial Agents

The most widely used method for aerosols capturing in indoor air environments is air filtration (Lee, 2011) due to their high efficiency and low costs (Xu and Yao, 2011). However, air filtration systems can become a source of contamination by harmful microbes. Antimicrobial filters inactivate the captured microbes and reduce the number of viable particles that get resuspended from the filter (Sim et al., 2015).

Different antimicrobial materials (silver, copper, carbon nanotubes) and biocidal chemicals coated on filters have shown promising results in bioaerosols inactivation (Sim et al., 2015). However, these inorganic nanoparticles can have negative impact on human health (Hwang et al., 2015a). After introduction into the human body through breathing or via skin, different organs can be affected (Sim et al., 2014b). Extracts of natural products (NPs) with antimicrobial activity have lately been considered as novel, efficient and cost–effective materials which can be used as filter media (Sim et al., 2015). They are typically less toxic compared to inorganic materials such as Ag, Cu and TiO_2 and more appropriate for indoor air filtration (Hwang et al., 2015b). Coating of filters with nano–sized NPs present an efficient method because they provide the maximum possible surface area in contact with the surrounding bioaerosols (Jung et al., 2011).

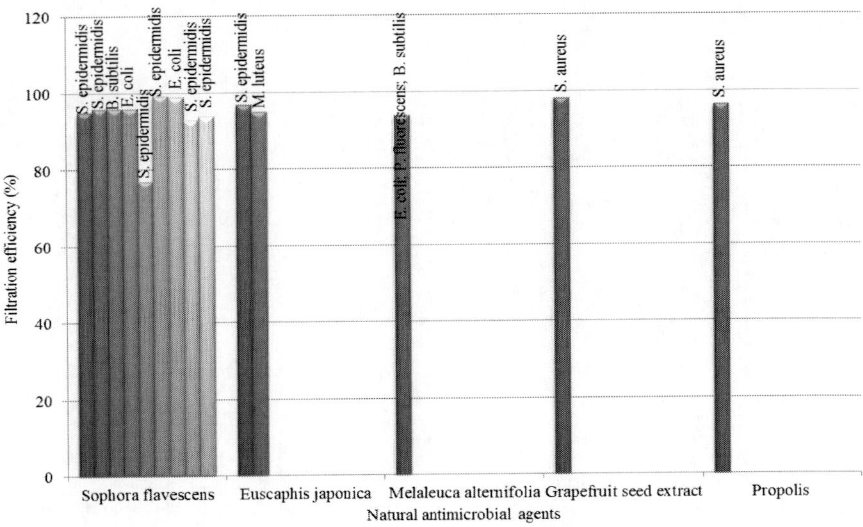

Figure 1. Filtration efficiency (%) of filters coated with different natural antimicrobial agents.

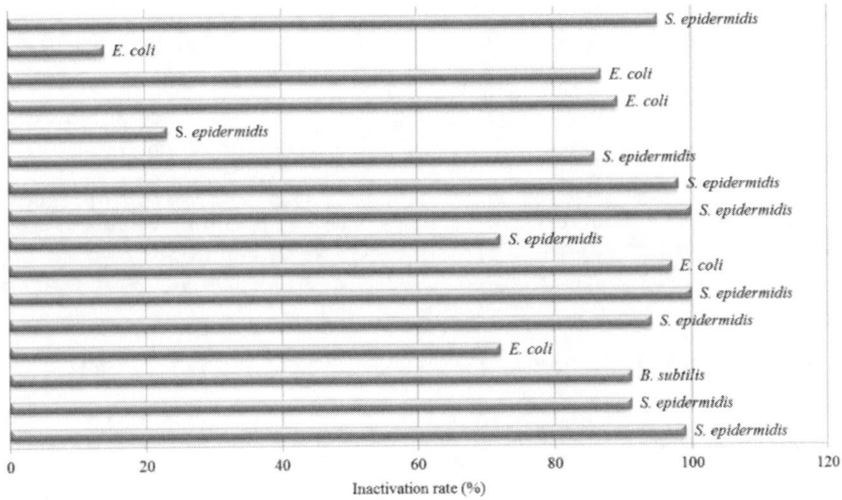

Figure 2. Inactivation rate (%) of bioaerosols by filters coated with Sophora flavescens nanoparticles.

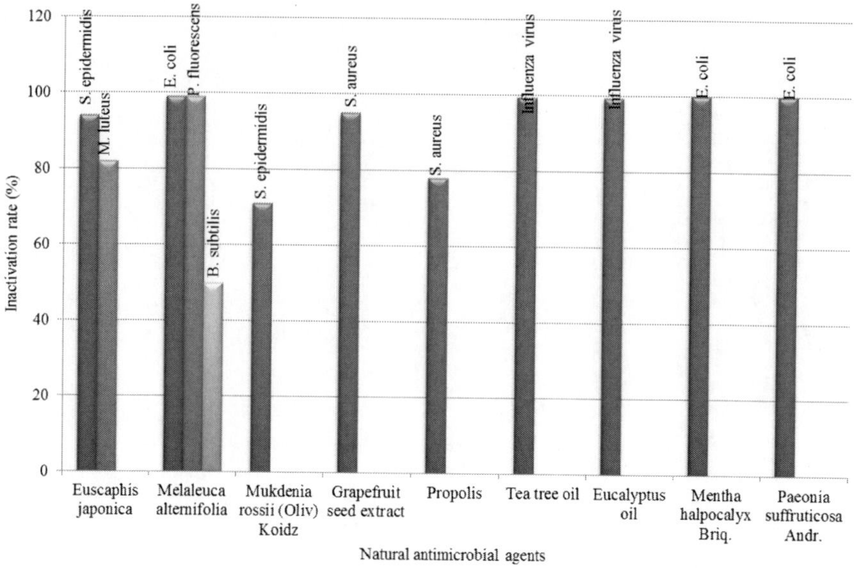

Figure 3. Inactivation rate (%) of bioaerosols by filters coated with natural antimicrobial agents.

The results of different studies (Sim et al., 2015; Hwang et al., 2015a; Jung et al., 2011; Sim et al., 2014a; Hwang et al., 2014; Sim et al., 2014b; Pyankov et al., 2008; Hwang et al., 2012; Jung et al., 2013; Lee et al., 2013; Han et al., 2015; Huang et al., 2010; Pyankov et al., 2012; Chong et al., 2013; Hwang et al., 2015b; Espiritu et al., 2014) of air filtration efficiency and inactivation rate of bioaerosols by filters coated with antimicrobial NPs are shown in Figures 1, 2 and 3.

5. Health Risks Associated with Bioaerosols

Human health can be affected with bioaerosols in many ways (Blais–Lecours et al., 2015). Inhalation, ingestion and skin contact are important routes of human exposure to airborne biological matter that could cause different health problems (Hurtado et al., 2014). Coughing, wheezing, runny nose, irritated eyes or throat, skin rash, diarrhea, aggravation of asthma, headache and fatigue are all health risks associated with being exposed to bioaerosols (Menetrez et al., 2009). However, the precise role of airborne biological matter in the development and aggravation of symptoms is poorly understood (Douwes et al., 2003).

Hypersensitivity, allergies and asthma are non–infectious diseases caused by bioerosols, while infectious diseases may be categorized into viral, bacterial and fungal diseases (Ghosh et al., 2015). Viruses are obligate parasites and airborne transmission is one of the major routes for viral infections spreading (Tseng and Li, 2005). Bacteria are considered as primary pathogens originating from animal sources, while plants represent sources for opportunistic pathogens (fungi) (Lacey and Dutkiewitcz, 1994). Gram–negative and Gram–positive bacteria make the airborne bacterial aerosols (Wei et al., 2014). Lipopolysaccharide (LPS) (Wei et al., 2014) and endotoxins (Schlosser et al., 2015) from the cell walls of Gram–negative bacteria may cause immune reaction. Infections of the upper respiratory tract (sinusitis, pharyngitis) are usually caused by viruses, while bacteria cause diseases of lower respiratory tract (bronchitis and pneumonia) (Blais–Lecours et al., 2015).

Aerobiological studies have shown that phyla Ascomycota and Basidiomycota make the majority of fungal spores in the outodoor air (Knutsen et al., 2012) and cause allergies (Haleem Khan and Mohan Karuppayil, 2012). Fungal metabolites may irritate respiratory system (Kalogerakis et al., 2005). 10% of people worldwide have fungal allergies (Jung et al., 2009a). *Aspergillus versicolor* and *Cladosporium cladosporioides* are usually found in indoor and outdoor environments (Lee et al., 2010a). Repeated exposure to *A. fumigatus* can cause allergic alveolitis and chronic bronchitis, while concentration of more than 10^6 CFU/m^3 is necessary to trigger an immune response (Deacon et al., 2009). *A. fumigatus* may cause sinusitis allergic alveolitis, asthma and allergic bronchopulmonary aspergillosis (Persoons et al., 2010). *Penicillium*, *Acremonium*, *Paecilomyces*, *Mucor* and *Cladosporium* cause respiratory infections and allergic reactions (Ghosh et al., 2015).

CONCLUSION

Microorganisms are ubiquitous in the environment; they are found in the air, soil, water, on plants, animals and humans. Major sources of bioaerosols are wastewater treatment plants, where microbes and non–viable biological matter become aerosolized during different mechanical and biological treatments.

Among different methods for indoor air purification, filtration is the most used technique. Special attention is given to filters that can simultaneously remove and inactivate bioaerosols and reduce health risks associated with microbial reintroduction into the air. Inorganic materials with antimicrobial activity are widely investigated, but they have negative impact on human health. New, efficient, cost–effective, less toxic (compared to inorganic materials) and environmentally friendly materials for filters coating are antimicrobial agents with natural origin.

Inactivation rate of nearly 100% of *E. coli* bioaerosols is achieved with filters coated with *Mentha haplocalyx* Briq. and *Paeonia suffruticosa* Andr. extracts. More than 99% of *S. epidermidis* bioaerosols were inactivated with *Sophora flavescens* nanoparticles coated on filters. Filters coated with tea tree oil and eucalyptus oil inactivated 99.5% of influenza virus, while grapefruit seed extract inactivates 95% of *S. epidermidis*. Filtration efficiency of 99% for *S. epidermidis* and *E. coli* show filters coated with *Sophora flavensis* nanoparticles.

ACKNOWLEDGMENTS

The financial support of the Ministry of Education, Science and Technological Development, Republic of Serbia, under theprojects III46010 and TR 33038, is acknowledged. We thank Professor Mara Manzalović from the Tachnical faculty in Bor (University of Belgrade) for providing language help.

REFERENCES

Abdel Hameed AA, Habeebuallah T, Mashat B, Elgendy S, Elmorsy TH, Elserougy S. Airborne fungal pollution at waste application facilities. *Aerobiologia* 2015;31:283–293.

Abdel Hameed AA, Khoder MI, Ibrahim YH, Saeed Y, Osman ME, Ghanem S. Study on some factors affecting survivability of airborne fungi. *Science of the Total Environment* 2012;414:696–700.

Abdel Hameed AA, Khoder MI, Yuosra S, Osman AM, Ghanem S. Diurnal distribution of airborne bacteria and fungi in the atmosphere of Helwan region, Egypt. *Science of the Total Environment* 2009;407:6217–6222.

Alagić SČ, Šerbula SM, Tošić SB, Pavlović AN, Petrović JV. Bioaccumulation of Arsenic and Cadmium in Birch and Lime from the Bor Region. *Archives of Environmental Contamination and Toxicology* 2013;65:671–682.

Antonijević MM, Dimitrijević MD, Stevanović ZO, Šerbula SM, Bogdanović GD. Investigation of the possibility of copper recovery from the flotation tailings by acid leaching. *Journal of Hazardous Materials* 2008;158:23–34.

Aríngoli EE, de la Luz Z. Basílico M, Altahus RL, Basílico JC. Multivariate analysis of fungal associations in the indoor air of Argentinean houses. *International Biodeterioration & Biodegradation* 2008;62:281–286.

Bauer H, Fuerhacker M, Zibuschka F, Schmid H, Puxbaum H. Bacteria and fungi in aerosols generated by two different types of wastewater treatment plants. *Water Research* 2002;36:3965–3970.

Blais–Lecours P, Perrott P, Duchaine C. Non–culturable bioaerosols in indoor settings: Impact on health and molecular approaches for detection. *Atmospheric Environment* 2015;110:45–53.

Brennen KP, Scarpino PV, Clark CS. Animal Viruses, Coliphages, and Bacteria in Aerosols and Wastewater at a Spray Irrigation Site. *Applied and Environmental Microbiology* 1988;54(2):409–415.

Breza–Boruta B, Paluszak Z. Influence of Water Treatment Plant on Microbiological Composition of Air Bioaerosols. *Polish Journal of Environmental Studies* 2007;16(5):663–670.

Bryan NC, Stewart M, Granger D, Guzik TG, Christner BC. A method for sampling microbial aerosols using high altitude balloons. *Journal of Microbiological Methods* 2014;107:161–168.

Chen X, Schluesener HJ. Nanosilver: A nanoproduct in medical application. *Toxicology Letters* 2008;176:1–12.

Chong E–s, Hwang GB, Nho CW, Kwon BM, Lee JE, Seo SC, et al., Antimicrobial durability of air filters coated with airborne *Sophora flavescens* nanoparticles. *Science of the Total Environment* 2013;444:110–114.

Chuaybamroong P, Chotigawin R, Supothina S, Sribenjalux P, Larpkiattaworn S, Wu C–Y. Efficacy of photocatalytic HEPA filter on microorganism removal. *Indoor Air* 2010;20:246–254.

Chuaybamroong P, Thunyasirinon C, Supothina S, Sribenjalux P, Wu C–Y. Performance of photocatalytic lamps on reduction of culturable airborne microorganism concentration. *Chemosphere* 2011;83:730–735.

Deacon LJ, Pankhurst LJ, Drew GH, Hayes ET, Jackson S, Longhurst PJ, et al., Particle size distribution of airborne *Aspergillus fumigatus* spores emitted from compost using membrane filtration. *Atmospheric Environment* 2009;43:5698–5701.

Ding W, Li L, Han Y, Liu J, Liu J. Site–related and seasonal variation of bioaerosol emission in an indoor wastewater treatment station: level, characteristics of particle size, and microbial structure. *Aerobiologia* 2015;1–14.

Douwes J, Thorne P, Pearce N, Heederik D. Bioaerosol Health Effects and Exposure Assessment: Progress and Prospects. *The Annals of Occupational Hygiene* 2003;47(3):187–200.

Drew GH, Jordinson GM, Smith MA, Pollard SJT. Evaluating the quality of bioaerosol risk assessments for composting facilities in England and Wales. *Resources, Conservation and Recycling* 2009;53:507–512.

Espiritu AG, Doma BT, Wang Y–F, Camille F, Masim P. Efficacy of methanol extracts of *Mentha haplocalyx* Briq. and *Paeonia suffruticosa* Andr. for potential antibacterial activity. *Sustainable Environment Research* 2014;24(5):319–324.

Faure M, Gerardin F, André J–C, Pons M–N, Zahraa O. Study of photocatalytic damages induced on *E. coli* by different photocatalytic supports (various types and TiO_2 configurations). *Journal of Photochemistry and Photobiology A: Chemistry* 2011;222:323–329.

Filipkowska Z, Janczukowitcz W, Krzemieniewski M, Pesta J. Microbiological Air Pollution in the Surroundings of the Wastewater Treatment Plant with Activated–Sludge Tanks Aerated by Horizontal Rotors. *Polish Journal of Environmental Studies* 2000;9(4):273–280.

Filipkowska Z, Janczukowitcz W, Krzemieniewski M, Pesta J. Municipal Wastewater Treatment Plant with Activated Sludge Tanks Aerated by CELPOX Devices as a Source of Microbiological Pollution of the Atmosphere. *Polish Journal of Environmental Studies* 2002;11(6):639–648.

Fracchia L, Pietronave S, Rinaldi M, Martinotti MG. Site–related airborne biological hazard and seasonal variations in two wastewater treatment plants. *Water Research* 2006a;40:1985–1994.

Fracchia L, Pietronave S, Rinaldi M, Martinotti MG. The assessment of airborne bacterial contamination in three composting plants revealed site–related biological hazard and seasonal variations. *Journal of Applied Microbiology* 2006b;100:973–984.

Galès A, Bru–Adan V, Godon J–J, Delabre K, Catala P, Ponthieux A, et al., Predominance of single bacterial cells in composting bioaerosols. *Atmospheric Environment* 2015;107:225–232.

Ghosh B, Lal H, Srivastava A. Review of bioaerosols in indoor environment with special reference to sampling, analysis and control mechanisms. *Environment International* 2015;85:254–272.

Gotkowska–Płachta A, Filipkowska Z, Korzeniewska E, Janczukowicz W, Dixon B, Gołaś I, et al., Airborne Microorganisms Emitted from Wastewater Treatment Plant Treating Domestic Wastewater and Meat Processing Industry Wastes. *Clean – Soil, Air, Water* 2013;41(5):429–436.

Green CF, Scarpino PV. The use of ultraviolet germicidal irradiation (UVGI) in disinfection of airborne bacteria. *Environmental Engineering and Policy* 2002;3:101–107.

Grinshpun SA, Adhikari A, Honda T, Kim KY, Toivola M, Rao KSR, Reponen T. Control of Aerosol Contaminants in Indoor Air: Combining the Particle Concentration Reduction with Microbial Inactivation. *Environmental Science and Technology* 2007;41:606–612.

Grinshpun SA, Adhikari A, Li C, Reponen T, Yermakov M, Schoenitz M, et al., Thermal inactivation of airborne viable *Bacillus subtilis* spores by short–term exposure in axially heated air flow. *Journal of Aerosol Science* 2010;41:352–363.

Grisoli P, Rodolfi M, Villani S, Grignani E, Cottica D, Berri A, et al., Assessment of airborne microorganism contamination in an industrial area characterized by an open composting facility and a wastewater treatment plant. *Environmental Research* 2009;109:135–142.

Haas D, Habib J, Luxner J, Galler H, Zarfel G, Schlacher R, et al., Comparison of background levels of culturable fungal spore concentrations in indoor and outdoor air in southeastern Austria. *Atmospheric Environment* 2014;98:640–647.

Haas D, Unteregger M, Habib J, Galler H, Marth E, Reinthaler FF. Exposure to Bioaerosol from Sewage Systems. *Water, Air, & Soil Pollution* 2010;207:49–56.

Haleem Khan AA, Mohan Karuppayil S. Fungal pollution of indoor environments and its management. *Saudi Journal of Biological Sciences* 2012;19:405–426.

Han B, Kang J–S, Kim H–J, Woo C–G, Kim Y–J. Investigation of Antimicrobial Activity of Grapefruit Seed Extract and Its Application to Air Filters with Comparison to Propolis and Shiitake. *Aerosol and Air Quality Research* 2015;15:1035–1044.

Han Y, Li L, Liu J, Zhang M. Microbial structure and chemical components of aerosols caused by rotating brushes in a wastewater treatment plant. *Environmental Science and Pollution Research* 2012;19:4097–4108.

Heikkinen MSA, Hjelmroos–Koski MK, Häggblom MM, Macher JM. Bioaerosol. In: Ruzer LS, Harley NH, editors. *Aerosols Handbook. Boca Raton: CRC Press;* 2005. p. 296–347.

Heinonen–Tanski H, Reponen T, Koivunen J. Airborne enteric coliphages and bacteria in sewage treatment plants. *Water Research* 2009;43:2558–2566.

Henningson EW, Ahlberg MS. Evaluation of Microbiological Aerosol Samplers: A Review. *Journal of Aerosol Science* 1994;25(8):1459–1492.

Heo KJ, Kim HB, Lee BU. Concentration of environmental fungal and bacterial bioaerosols during the monsoon season. *Journal of Aerosol Science* 2014;77:31–37.

Huang C–Y, Lee C–C, Li F–C, Ma Y–P, Su H–JJ. The seasonal distribution of bioaerosols in municipal landfill sites: a 3–yr study. *Atmospheric Environment* 2002;3:4385–4395.

Huang R, Agranovski I, Pyankov O, Grinshpun S. Removal of viable bioaerosol particles with low–efficiency HVAC filer enhanced by continuous emission of unipolar air ions. *Indoor Air* 2008;18:106–112.

Huang R, Pyankov OV, Yu B, Agranovski IE. Inactivation of Fungal Spores Collected on Fibrous Filters by *Melaleuca alternifolia* (Tea Tree Oil). *Aerosol Science and Technology* 2010;44:262–268.

Hurtado L, Rodríguez G, López J, Castillo JE, Milina L, Zavala M, et al., Characterization of atmospheric bioaerosols at 9 sites in Tijuana, Mexico. *Atmospheric Environment* 2014;96:430–436.

Hwang GB, Heo KJ, Yun JH, Lee JE, Lee HJ, Nho CW, et al., Antimicrobial Air Filters Using Natural *Euscaphis japonica* Nanoparticles. *PLoS ONE* 2015a;10(5):1–14.

Hwang GB, Jung JH, Jeong TG, Lee BU. Effect of hybrid UV–thermal energy stimuli on inactivation of *S. epidermidis* and *B. subtilis* bacterial bioaerosols. *Science of the Total Environment* 2010;408:5903–5909.

Hwang GB, Lee JE, Nho CW, Lee BU, Lee SJ, Jung JH, et al., Short–term effect of humid airflow on antimicrobial air filters using *Sophora flavescens* nanoparticles. *Science of the Total Environment* 2012;421–422:273–279.

Hwang GB, Park H–S, Bae G–N, Jung JH. Effects of Electric Field Strength on an Antimicrobial Air Filter. *Aerosol and Air Quality Research* 2014;14:1028–1037.

Hwang GB, Sim KM, Bae G–N, Jung JH. Synthesis of hybrid carbon nanotube structures coated with *Sophora flavescens* nanoparticles and their application to antimicrobial air filtration. *Journal of Aerosol Science* 2015b;89:44–54.

Joe YH, Woo K, Hwang J. Fabrication of an anti–viral air filter with SiO_2–Ag nanoparticles and performance evaluation in a continuous airflow condition. *Journal of Hazardous Materials* 2014;280:356–363.

Johansson E, Adhikari A, Reponen T, Yermakov M, Grinshpun SA. Association Between Increased DNA Mutational Frequency and Thermal Inactivation of Aerosolized *Bacillus* Spores Exposed to Dry Heat. *Aerosol Science and Technology* 2011;45:376–381.

Jones AM, Harrison RM. The effects of meteorological factors on atmospheric bioaerosol concentrations–a review. *Science of the Total Environment* 2004;326:151–180.

Jung JH, Hwang GB, Park SY, Lee JE, Nho CW, Lee BU, et al., Antimicrobial Air Filtration Using Airborne *Sophora Flavescens* Natural–Product Nanoparticles. *Aerosol Science and Technology* 2011;45:1510–1518.

Jung JH, Lee CH, Lee JE, Lee JH, Kim SS, Lee BU. Design and characterization of a fungal bioaerosol generator using multi–orifice air jets and a rotating substrate. *Journal of Aerosol Science* 2009a(1);40:72–80.

Jung JH, Lee JE, Bae G–N. Use of electrosprayed *Sophora flavescens* natural–product nanoparticles for antimicrobial air filtration. *Journal of Aerosol Science* 2013;57:185–193.

Jung JH, Lee JE, Kim SS. Thermal effects on bacterial bioaerosols in continuous air flow. *Science of the Total Environment* 2009c;407:4723–4730.

Jung JH, Lee JE, Lee CH, Kim SS, Lee BU. Treatment of Fungal Bioaerosols by a High–Temperature, Short–Time Process in a Continuous–Flow System. *Applied and Environmental Microbiology* 2009b;75(9):2742–2749.

Kacprzak M, Naczaj E, Okoniewska E. The comparative mycological analysis of wastewater and sewage sludges from selected wastewater treatment plants. *Desalination* 2005;185:363–370.

Kalinović TS, Šerbula SM, Radojević AA, Kalinović JV, Steharnik MM, Petrović JV. Elder, linden and pine biomonitoring ability of pollution emitted from the copper smelter and the tailings ponds. *Geoderma* 2016;262:266–275.

Kalogerakis N, Paschali D, Lekaditis V, Pantidou A, Eleftheriadis K, Lazaridis M. Indoor air quality–bioaerosol measurements in domestic and office premises. *Journal of Aerosol Science* 2005;36(5–6):751–761.

Karra S, Katsivela E. Microorganisms in bioaerosol emissions from wastewater treatment plants during summer at a Mediterranean site. *Water Research* 2007;41:1355–1365.

Keller V, Keller N, Ledoux MJ, Lett M–C. Biological agent inactivation in a flowing air stream by photocatalysis. *Chemical Communications* 2005;23:2918–2920.

Kim YS, Yoon KY, Park JH, Hwang J. Application of air ions for bacterial de–colonization in air filters contaminated by aerosolized bacteria. *Science of the Total Environment* 2011;409:748–755.

Knutsen AP, Bush RK, Demain JG, Denning DW, Dixit A, Fairs A, et al., Fungi and allergic lower respiratory tract diseases. *Journal of Allergy and Clinical Immunology* 2012;129(2):280–291.

Ko G, First MW, Burge HA. The Characterization of Upper–Room Ultraviolet Germicidal Irradiation in Inactivating Airborne Microorganisms. *Environmental Health Perspectives* 2002;110(1):95–101.

Korzeniewska E, Filipkowska Z, Gotkowska–Płachta A, Janczukowicz W, Dixon B, Czułowska M. Determination of emitted airborne microorganisms from a BIO–PAK wastewater treatment plant. *Water Research* 2009;43:2841–2851.

Korzeniewska E, Harnisz M. Culture–Dependent and Culture–Independent Methods in Evaluation of Emission of *Enterobacteriaceae* from Sewage to the Air and Surface Water. *Water, Air, & Soil Pollution* 2012;223:4039–4046.

Kruczalak K, Olańczuk–Neyman K. Microorganisms in the Air Over Wastewater Treatment Plants. *Polish Journal of Environmental Studies* 2004;13(5):537–542.

Kujundzic E, Hernandez M, Miller SL. Ultraviolet germicidal irradiation inactivation of airborne fungal spores and bacteria in upper–room air and HVAC in–duct configurations. *Journal of Environmental Engineering and Science* 2007;6:1–9.

Kujundzic E, Matalkah F, Howard CJ, Hernandez M, Miller SL. UV Air Cleaners and Upper–Room Air Ultraviolet Germicidal Irradiation for Controlling Airborne Bacteria and Fungal Spores. *Journal of Occupational and Environmental Hygiene* 2006,3:536–546.

Lacey J, Dutkiewicz J. Bioaerosols and Occupational Lung Disease. *Journal of Aerosol Science* 1994;25(8):1371–1404.

Lavoie J, Dunkerley CJ, Kosatsky T, Dufresne A. Exposure to aerosolized bacteria and fungi among collectors of commercial, mixed residential, recyclable and compostable waste. *Science of the Total Environment* 2006;370:23–28.

Lee BU, Jung JH, Yun SH, Hwang GB, Bae GN. Application of UVAPS to real–time detection of inactivation of fungal bioaerosols due to thermal energy. *Journal of Aerosol Science* 2010b;41:694–701.

Lee BU, Yun SH, Ji J–H, Bae G–N. Inactivation of *S. epidermidis*, *B. subtilis*, and *E. coli* Bacteria Bioaerosols Deposited on a Filter Utilizing Airborne Silver Nanoparticles. *Journal of Microbiology and Biotechnology* 2008;18(1):176–182.

Lee BU. Life Comes from the Air: A Short Review on Bioaerosol Control. *Aerosol and Air Quality Research* 2011;11:921–927.

Lee DH, Jung JH, Lee BU. Effect of Treatment with a Natural Extract of *Mukdenia Rossii (Oliv) Koidz*and Unipolar Ion Emission on the Antibacterial Performance of Air Filters. *Aerosol and Air Quality Research* 2013;13:771–776.

Lee JH, Hwang GB, Jung JH, Lee DH, Lee BU. Generation characteristics of fungal spore and fragment bioaerosols by airflow control over fungal cultures. *Journal of Aerosol Science* 2010a;41:319–325.

Lee S–G, Hyun J, Lee SH, Hwang J. One–pass antibacterial efficacy of bipolar air ions against aerosolized *Staphyloccocus epidermidis* in a duct flow. *Journal of Aerosol Science* 2014;69:71–81.

Lehtinen J, Tolvanen O, Nivukoski U, Veijanen A, Hänninen K. Occupational hygiene in terms of volatile organic compounds (VOCs) and bioaerosols at two solid waste management plants in Finland. *Waste Management* 2013:33;964–973.

Li J, Zhou L, Zhang X, Xu C, Dong L, Yao M. Bioaerosol emissions and detection of airborne antibiotic resistance genes from a wastewater treatment plant. *Atmospheric Environment* 2016:124:404–412.

Li L, Gao M, Liu J. Distribution characterization of microbial aerosols emitted from a wastewater treatment plant using the Orbal oxidation ditch process. *Process Biochemistry* 2011;46:910–915.

Li L, Hah M, Liu J. Assessing genetic structure, diversity of bacterial aerosol from aeration system in an oxidation ditch wastewater treatment plant by culture methods and bio–molecular tools. *Environmental Monitoring and Assessment* 2013c;185:603–613.

Li Y, Qiu X, Li M, Ma Z, Niu T, Feng Y. Concentration and Size Distribution of Airborne Actinomycetes in a Municipal Wastewater Treatment Plant. *Polish Journal of Environmental Studies* 2012;21(5):1305–1311.

Li Y, Yang L, Meng Q, Qiu X, Feng Y. Emission Characteristics of Microbial Aerosol in a Municipal Sewage Treatment Plant in Xi'an, China. *Aerosol and Air Quality Research* 2013b;13:343–349.

Li Y, Zhang H, Qiu X, Zhang Y, Wang H. Dispersion and Risk Assessment of Bacterial Aerosols Emitted from Rotating–Brush Aerator during Summer in a Wastewater Treatment Plant of Xi'an, China. *Aerosol and Air Quality Research* 2013a;13:1807–1814.

Lin C–Y, Li C–S. Control Effectiveness of Ultraviolet Germicidal Irradiation on Bioaerosols. *Aerosol Science and Technology* 2002;36:474–478.

Maharia S, Srivastava A. Influence of seasonal variation on concentration of fungal bioaerosol at a sewage treatment plant (STP) in Delhi. *Aerobiologia* 2015;31:249–260.

Małecka–Adamowicz M, Donderski W, Dokładna W. Microflora of Air in the Sewage Treatment Plant of Kapuściska in Bydgoszcz. *Polish Journal of Environmental Studies* 2011;20(5):1235–1242.

Marambio–Jones C, Hoek EMV. A review on the antibacterial effects of silver nanomaterials and potential applications for human health and the environment. *Journal of Nanoparticle Research* 2010;12:1531–1551.

Marcó MB, del LujánQuiberoni A, Negro AC, Reinheimer JA, Alfano OM. Evaluation of the photocatalytic inactivation efficiency of dairy bacteriophages. *Chemical Engineering Journal* 2011;172:987–993.

Masclaux FG, Hotz P, Friedli D, Savova–Bianchi D, Oppilager A. High occurrence of hepatitis E virus in samples from wastewater treatment plants in Switzerland and comparison with other enteric viruses. *Water Research* 2013;47:5101–5109.

Menetrez MY, Foarde KK, Esch RK, Schwartz TD, Dean TR, Hays MD, et al., An evaluation of indoor and outdoor biological particulate matter. *Atmospheric Environment* 2009;43:5476–5483.

Michałkiewicz M, Pruss A, Dymaczewski Z, Jeż–Walkowiak J, Kwaśna S. Microbiological Air Monitoring around Municipal Wastewater Treatment Plants. *Polish Journal of Environmental Studies* 2011;20(5):1243–1250.

Miller SL, Macher JM. Evaluation of a Methodology for Quantifying the Effect of Room Air Ultraviolet Germicidal Irradiation on Airborne Bacteria. *Aerosol Science and Technology* 2000;33:274–295.

Nasir ZA, Colbeck I, Sultan S, Ahmed S. Bioaerosols in residential micro–environments in low income countries: A case study from Pakistan. *Environmental Pollution* 2012;168:15–22.

Niazi S, Hassanvand MS, Mahvi AH, Nabizadeh R, Alimohammadi M, Nabavi S, et al., Assessment of bioaerosol contamination (bacteria and fungi) in the largest urban wastewater plant in the Middle East. *Environmental Science and Pollution Research* 2015;22:16014–16021.

Oppliger A, Hilfiker S, Vu Duc T. Influence of Seasons and Sampling Strategy on Assessment of Bioaerosols in Sewage Treatment Plants in Switzerland. *The Annals of Occupational Hygiene* 2005;49(5):393–400.

Orsini M, Laurenti P, Boninti F, Arzani D, Ianni A, Romano–Spica V. A molecular typing approach for evaluating bioaerosol exposure in wastewater treatment plants workers. *Water Research* 2002;36:1375–1378.

Pankhurst LJ, Akeel U, Hewson C, Maduka I, Pham P, Saragossi J, et al., Understanding and mitigating the challenge of bioaerosol emissions from urban community composting. *Atmospheric Environment* 2011a;45:85–93.

Pankhurst LJ, Deacon LJ, Liu J, Drew GH, Hayes ET, Jackson S, et al., Spacial variations in airborne microorganism and endotoxin concentrations at green waste composting facilities. *International Journal of Hygiene and Environmental Health* 2011b;214:376–383.

Park J–H, Yoon K–Y, Kim Y–S, Byeon JH, Hwang J. Removal of submicron aerosol particles and bioaerosols using carbon fiber ionizer assisted fibrous medium filter media. *Journal of Mechanical Science and Technology* 2009;23:1846–1851.

Pascual L, Pérez–Luz S, Yáñez MA, Santamaría A, Gilbert K, Salgot M, et al., Bioaerosol emission from wastewater treatment plants. *Aerobiologia* 2003;19:261–270.

Persoons R, Parat S, Stoklov M, Perdrix A, Maitre A. Critical working tasks and determinants of exposure to bioaerosols and MVOC at composting facilities. *International Journal of Hygiene and Environmental Health* 2010;213:338–347.

Pham T–D, Lee B–K. Effects of Ag doping on the photocatalytic disinfection of *E. coli* in bioaerosol by Ag–TiO_2/GF under visible light. *Journal of Colloid and Interface Science* 2014a;428:24–31.

Pham T–D, Lee B–K. Feasibility of Silver Doped TiO_2/Glass Fiber Photocatalyst under Visible Irradiation as an Indoor Air Germicide.

International Journal of Environmental Research and Public Health 2014b;11:3271–3288.

Pham T–D, Lee B–K. Novel integrated approach of adsorption and photo–oxidation using Ag–TiO_2/PU for bioearosol removal under visible light. *Chemical Engineering Journal* 2015;275:357–365.

Pigeot–Remy S, Real P, Simonet F, Hernandez C, Vallet C, Lazzaroni JC, et al., Inactivation of *Aspergillus niger* spores from indoor air by photocatalytic filters. *Applied Catalysis B: Environmental* 2013;134–135:167–173.

Polymenakou PN. Atmosphere: A source of Pathogenic or Beneficial Microbes? *Atmosphere* 2012;3:87–102.

Prażmo Z, Krysińska–Traczyk E, Skórska C, Sitkowska J, Cholewa G, Dutkiewicz. Exposure to Bioaerosols in a Municipal Sewage Treatment Plant. *Annals of Agricultural and Environmental Medicine* 2003;10:241–248.

Pyankov OV, Agranovski IE, Huang R, Mullins BJ. Removal of Biological Aerosols by Oil Coated Filters. *Clean – Soil, Air, Water* 2008;36(7):609–614.

Pyankov OV, Usachev EV, Pyankova O, Agranovski IE. Inactivation of Airborne Influenza Virus by Tea Tree and Eucalyptus Oils. *Aerosol Science and Technology* 2012;46:1295–1302.

Rahkonen P. Airborne Contaminants at Waste Treatment Plants. *Waste Management & Research* 1992;10:411–421.

Sánchez–Monedero MA, Aguilar MI, Fenoll R, Roig A. Effect of the aeration system on the levels of airborne microorganisms generated at wastewater treatment plants. *Water Research* 2008;42:3739–3744.

Sánchez–Monedero MA, Stentiford EI, Urpilainen ST. Bioaerosol Generation at Large–Scale Green Waste Composting Plants. *Journal of the Air & Waste Management Association* 2005;55:612–618.

Sánchez–Monedero MA, Stentiford EI. Generation and Dispersion of Airborne Microorganisms from Composting Facilities. *Trans IChemE* 2003;81:166–170.

Schlosser O, Déportez IZ, Facon B, Fromont E. Extension of the sorting instructions for household plastic packaging and changes in exposure to bioaerosols at materials recovery facilities. *Waste Management* 2015;46:47–55.

Šerbula SM, Antonijević MM, Milosević NM, Milić SM, Ilić AA. Concentrations of particulate matter and arsenic in Bor (Serbia). *Journal of Hazardous Materials* 2010;181:43–51.

Šerbula SM, Stevanović J, Trujić V. Arsenic, Heavy Metals and SO_2 Derived in a Mining–Metalurgical Production Process. In: Satinder Kaur Brar, editor. Hazardous Materials: Types, Risks and Control. New York: *Nova Science Publishers US;* 2011. p. 187–223.

Šerbula SM, Miljković DĐ, Kovacević RM, Ilić AA. Assessment of airborne heavy metal pollution using plant parts and topsoil. *Ecotoxicology and Environmental Safety* 2012a;76:209–214.

Šerbula SM, Alagić SČ, Ilić AA, Kalinović TS, Strojić JV. Particulate Matter Originated From Mining–Metallurgical Processes. In: Henrik Knudsen, Niels Rasmussen, editors. Particulate Matter: Sources, Emission Rates and Health Effects. New York: *Nova Science Publishers US;* 2012b. p. 91–116.

Šerbula SM, Kalinović TS, Ilić AA, Kalinović JV, Steharnik MM. Assessment of airborne heavy metal pollution using Pinus spp. and Tilia spp., *Aerosol and Air Quality Research,* 2013a;13:563–573.

Šerbula S, Radulović M, Zafirović I, Stanižan Z, Petrović M, Vukelić J, Petrović I, Vesković M, *Environmental protection in Pančevo and Bor: challenges of participatory approach to environmental governance,* University of Belgrade, Faculty of Philosophy, Institute for Sociological Research, Belgrade, First Edition 2013b. (ISBN 978–86–913927–5–8)(CIP 502/504(497.11)(082))(COBISS.SR–ID 197486092) Chapter: Šerbula Snežana: Environmental Pollution in the Bor Region, 14–29.

Šerbula SM, Kalinović TS, Kalinović JV, Ilić AA. Exceedance of air quality standards resulting from pyro–metallurgical production of copper: a case study, *Bor (Eastern Serbia), Environmental Earth Sciences* 2013c;68: 1989–1998.

Šerbula SM, Radojević AA., Kalinović JV, Kalinović TS. Indication of airborne pollution by birch and spruce in the vicinity of copper smelter. *Environmental Science and Pollution Research 2014a;21:11510–11520.*

Šerbula SM, Ilić AA, Kalinović JV, Kalinović TS, Petrović NB. Assessment of air pollution originating from copper smelter in Bor (Serbia), *Environmental Earth Sciences* 2014b;71:1651–1661.

Šerbula SM, Živković DT, Radojević AA, Kalinović TS, Kalinović JV. Emission of SO_2 And SO_4^{2-} from Copper Smelter and its Influence on the Level of Total S in Soil and Moss in Bor and the Surroundings, *Hemijska industrija* 2015;69(1):51–58.

Šerbula S, Stanković V, Živković D, Kamberović Ž, Gorgievski M, Kalinović T: Characteristics of Wastewater Streams within the Bor Copper Mine and their Influence on Pollution of the Timok River, Serbia. *Mine Water and the Environment* 2016;32(104):1–6.

Sharma VK, Yngard RA, Lin Y. Silver nanoparticles: Green synthesis and their antimicrobial activities. *Advances in Colloid and Interface Science* 2009;145:83–96.

Sialve B, Gales A, Hamelin J, Wery N, Steyer J–P. Bioaerosol emissions from open microalgal processes and their potential environmental impacts: what can be learned from natural and anthropogenic aquatic environments? *Current Opinion in Biotechnology* 2015;33:279–786.

Sim KM, Kim KH, Hwang GB, Seo SC, Bae G–N, Jung JH. Development and evaluation of antimicrobial activated carbon fiber filters using *Sophora flavescens* nanoparticles. *Science of the Total Environment* 2014a;493:291–297.

Sim KM, Lee HJ, Nho CW, Bae G–N, Jung JH. Effects of Surrounding Temperature on Antimicrobial Air Filters Coated with *Sophora flavescens* Nanoparticles. *Aerosol Science and Technology* 2014b;48:324–332.

Sim KM, Park H–S, Bae G–N, Jung JH. Antimicrobial nanoparticle–coated electrostatic air filter with high filtration efficiency and low pressure drop. *Science of the Total Environment* 2015;533:266–274.

Solans X, Alonso RM, Constans A, Mansilla A. Exposición laboral a hongos y bacteria sambientales en la plata de selección de residuos de envases. *Revista Iberoamericana de Micología* 2007;24:131–135.

Stellacci P, Libreti L, Notarnicola M, Haas CN. Hygienic sustainability of site location of wastewater treatment plants. A case study. II. Estimating airborne biological hazard. *Desalination* 2010;253:106–111.

Sykes P, Morris RHK, Allen JA, Wildsmith JD, Jones KP. Workers' exposure to dust, endotoxin and β–(1–3) glucan at four large–scale composting facilities. *Waste Management* 2011;31:423–430.

Taha MPM, Drew GH, Longhurst PJ, Smith R, Pollard SJT. Bioaerosol releases from compost facilities: Evaluating passive and active source terms at a green waste facility for improves risk assessments. *Atmospheric Environment* 2006;40:1159–1169.

Taha MPM, Drew GH, Tamer A, Hewings G, Jordinson GM, Longhurts PJ, et al., Improving bioaerosol exposure assessments of composting facilities – Comparative modelling of emissions from different compost ages and processing activities. *Atmospheric Environment* 2007;41:4504–4519.

Taha MPM, Pollard SJT, Sarkar U, Longhurst P. Estimating fugitive bioaerosol releases from static compost windrows: Feasibility of a portable wind tunnel approach. *Waste Management* 2005;25:445–450.

Tamer Vestlund A, Al–Ashaab R, Tyrrel SF, Longhurst PJ, Pollard SJT, Drew GH. Morphological classification of bioaerosols from composting using scanning electron microscopy. *Waste Management* 2014;34:1101–1108.

Tseng C–C, Li C–S. Collection efficiencies of aerosol samplers for virus–containing aerosols. *Aerosol Science* 2005;36:593–607.

Uhrbrand K, Schultz AC, Madsen AM. Exposure to Airborne Noroviruses and Other Bioaerosol Components at a Wastewater Treatment Plant in Denmark. *Food and Environmental Virology* 2011;3:130–137.

Van Leuken JPG, Swart AN, Havelaar AH, Van Pul A, Van der Hoek W, Heederik. Atmospheric dispersion modelling of bioaerosols that are pathogenic to humans and livestock – A review to inform risk assessment studies. *Microbial Risk Analysis.* 2016;1:19–39.

Vilavert L, Nadal M, Inza I, Figueras MJ, Domingo JL. Baseline levels of bioaerosols and volatile organic compounds around a municipal waste incinerator prior to the construction of a mechanical–biological treatment plant. *Waste Management* 2009;29:2454–2461.

Vítězová M, Vítěz T, Mlejnková H, Lošák T. Microbial Contamination of the Air at the Wastewater Treatment Plant. *Acta Universitatis Agriculturae Et Silviculturae Mendelianae Brunensis* 2012;LX(3):233–240.

Wang W, Ma Y, Ma X, Wu f, Ma X, An L, et al., Seasonal variations of airborne bacteria in the Mogao Grottoes, Dunhuang, China. *International Biodeterioration & Biodegradation* 2010;64:309–315.

Wei K, Zou Z, Yao M. Charge levels and Gram (±) fractions of environmental bacterial aerosols. *Journal of Aerosol Science* 2014;74:52–62.

Woo AC, Brar MS, Chan Y, Lau MCY, Lueng FCC, Scott JA, et al., Temporal variation in airborne microbial populations and microbially–derived allergens in a tropical urban landscape. *Atmospheric Environment* 2013;74:291–300.

Woo M–H, GrippinA, Wu C–Y, Wander J. Microwave–irradiation–assisted HVAC filtration for Inactivation of Viral Aerosols. *Aerosol and Air Quality Research* 2012;12:295–303.

Wu Y, Yao M. Effects of microwave irradiation on concentration, diversity and gene mutation of culturable airborne microorganisms of inhalable sizes in different environments. *Journal of Aerosol Science* 2011;42:800–810.

Wu Y, Yao M. Inactivation of bacteria and fungus aerosols using microwave irradiation. *Journal of Aerosol Science* 2010;41:682–693.

Xu P, Peccia J, Fabian P, Martyny JW, Fennelly KP, Hernandez M, et al., Efficacy of ultraviolet germicidal irradiation of irradiation of upper–room air in inactivating airborne bacterial spores and mycobacteria in full–scale studies. *Atmospheric Environment* 2003;37:405–419.

Xu Z, Yao M. Effects of single–walled carbon nanotube filter on culturability and diversity of environmental bioaerosols. *Journal of Aerosol Science* 2011;42:387–396.

Yadav J, Kumar A, Mahor P, Kumar Goel A, Singh Chaudrau H, Kumar Yadava P, et al., Distribution of airborne microbes and antibiotic susceptibility pattern of bacteria during Gwalior trade fair, Central India. *Journal of the Formosan Medical Association* 2015;114:639–646.

Yao N, Yeung KL. Investigation of the performance of TiO_2photocatalytic coatings. *Chemical Engineering Journal* 2011;167:13–21.

Yu K–P, Lee GW–M, Lin S–Y, Huang CP. Removal of bioaerosols by the combination of a photocatalytic filter and negative air ions. *Journal of Aerosol Science* 2008;39:377–392.

Zacarías SM, Satuf ML, Vaccari MC, Alfano OM. Photocatalytic inactivation of bacterial spores using TiO_2 films with silver deposits. *Chemical Engineering Journal* 2015;266:133–140.

Zhang Q, Damit B, Welch J, Park H, Wu C–Y, Sigmund W. Microwave assisted nanofibrous air filtration for disinfection of bioaerosols. *Journal of Aerosol Science* 2010;41:880–888.

In: Air Pollution
Editor: Gerald L. Burns

ISBN: 978-1-63485-374-3

Chapter 5

HEAVY METALS AIR POLLUTION STUDY IN MINES ENVIRONMENTS. CASE STUDY BREGALNICA RIVER BASIN, REPUBLIC OF MACEDONIA

Trajče Stafilov[1,*], Biljana Balabanova[2] and Jasminka Alijagić[3]

[1]Institute of Chemistry, Faculty of Science, Ss. Cyril and Methodius University, Skopje, R. Macedonia
[2]Faculty of Agriculture, Goce Delčev University, R. Macedonia
[3]Geological Survey of Slovenia, Ljubljana, Slovenia

ABSTRACT

Application of several moss species and attic dust for monitoring of anthropogenic impact on heavy metals air pollution in Bregalnica River Basin, Republic of Macedonia, was studied. Moss samples were reviewed for their potential to reflect heavy metals air pollution. The attention was focused on their quantification ability, underlying the metal accumulation within moss plant tissue and attic dust "historical archiving." Potential "hot spots" were selected in areas of copper mine (Bučim mine) and lead and zinc mines (Zletovo mine and Sasa mine) as main metal pollution sources in the Eastern part of the Republic of Macedonia. Continuously,

* Corresponding author: Email: trajcest@pmf.ukim.mk.

dust distribution from ore and flotation tailings occurs. This results with air-introduction and deposition of higher contents of certain metals. Several moss species (*Hypnum cupressiforme*, *Homalothecium lutescens* and *Scleropodium purum*) were used as plant sampling media. Determination of chemical elements was conducted by using both instrumental techniques: atomic emission spectrometry with inductively coupled plasma (ICP-AES) and mass spectrometry with inductively coupled plasma (ICP-MS). Combination of multivariate techniques (PCA, FA and CA) was applied for data processing and identification of elements association with lithogenic or anthropogenic origin. Spatial distribution maps were constructed for determination and localizing of narrower areas with higher contents of certain anthropogenic elements. In this way influences of selected human activities in local (small scale) air pollution can be determined. Summarized data reveal real quantification of the elements distribution not only in order determination of hazardously elements distribution, but also present complete characterization for elements deposition in mines environs.

Keywords: chemical elements, toxic metals, monitoring, moss, attic dust, air pollution, Bregalnica River Basin, Republic of Macedonia

1. Introduction

The environmental pollution at hazardously levels for living presents a global problem and a macro-case for monitoring. Some sub-disciplines were occurred with time in order to consider the realistic environmental conditions. Arguably, the understanding of atmospheric pollution is one of the most emergent areas of the environmental science (Fernández, 2015). Atmospheric pollution represents solutions or suspensions of minute amounts of harmful compounds in the air (Wang et al., 2009). The degree and the extent of environmental changes over the last decades has given a new urgency and relevance for detection and understanding of environmental changes, due to human activities, which have altered global biogeochemical cycling of heavy metals and other pollutants (Athar and Vohora, 1995; VanLoon and Duffy, 2000; Akimoto, 2003). Monitoring toxic air pollutants is needed for understanding their spatial and temporal distribution and ultimately to minimize their harmful effects. In addition, to direct physical and chemical methods of air pollution monitoring, *bioindication* has also been used to evaluate air pollution risk (Rühling and Tyler, 1971, 1973; Kanaroglou et al., 2005; Valero, 2008; Aboal et al., 2010).

Mosses have been frequently used to monitor time-integrated bulk deposition of metals as a combination of wet, cloud, and dry deposition, thus eliminating some of the complications of precipitation analysis due to the heterogeneity of precipitation (Coşkun et al., 2005; Tyler, 2008). Ectohydric mosses in particular draw negligible amounts of water and minerals from the soil, and almost entirely depend on the atmospheric inputs of nutrients (Ruhling and Tyler, 1971). Because mosses have a high cation exchange capacity (CEC), they act as hyper-accumulators of metals and metal complexes. The metals are bound to the tissue with minimal translocation within the plant due to a lack of vascular tissue (Aceto et al., 2003). This results in biological tissue that can be analyzed to reveal time-integrated deposition (Van het Bolcher et al., 2006). Additional advantages of using mosses as heavy metal biomonitors include their stationary nature, widespread geographical distribution, and low genetic variability between populations. It has been shown that there is some experimental error due to heterogeneity in morphological characteristics and microenvironments among different populations (Sokhi, 2011; Manara, 2012). There is also an incomplete understanding of the degree of mineral uptake by ectohydric mosses in direct contact with substrate (Marschner, 1995, Market, 2007). Despite the accuracy and precision of precipitation analysis techniques, however, mosses offer an efficient, low-cost complement for determining metal concentrations at a large number of locations and offer analyses of biologically relevant fluxes at multiple scales.

Harmens and his European colleagues have found that mosses are reliable indicators of air pollution risks to ecosystems; because they get most of their nutrients direct from the air and rain, rather than the soil. Since 2000, the European moss survey has been conducted by a special international program (ICP Vegetation). Moss data provides a better geographical coverage than measured deposition data and reveals more about actual atmospheric pollution at a local level (http://icpvegetation.ceh.ac.uk/). Latest data reported from Harmens et al., (2015) and Barandovski et al., (2013) indicates on the significant enrichments of some toxic elements.

An alternative approach is to use exposed attic dust, because over long periods of time atmospheric particles can accumulate, providing a record of historical local deposition (Alijagič and Šajn, 2001, 2006; Cizdziel et al., 2000; Ilacqua et al., 2003, Jeffrey et al., 2005; Gosar et al., 2006; Jemec Auflič and Šajn, 2007; Žibret, 2012; Žibret and Šajn, 2008a, 2008b, 2010; Šajn, 1999, 2000, 2001, 2002, 2003, 2005, 2006; Šajn et al., 2012). The determination of historical emissions is based on the data of heavy metals concentration in the attic dust from different measurement sites of the weight of total monthly air deposit. The main idea behind determining past emissions is that heavy metals characterization of deposited dust on a small area is multiplied by the concentration of the elements in that area, the mass of the polluter which has been transported to the place of interest by air can be determinate. Undisturbed attic dusts are potential archive for atmospherically deposited particles and have been shown to be effective across urban areas and in the vicinity of smelters, mines and other potentially emission sources. These have the additional benefit that they are protected from wash and aqueous alteration that may affect other tips of samples (monitors) over long periods of time. Atmospheric emissions attributed to the extraction stage of mining come mainly from the action of wind on disturbed land and stockpiles of ore and waste material. As a result of these processes in the atmosphere permanently introduces dust (Sengupta 1993, Ilacqua et al., 2003). Continuous monitoring of the content of heavy metals and other toxic components contained in dust is possible thought implementation of monitoring by using samples of dust from attic beams (Ilacqua et al., 2003, Atriola et al., 2004). Attic dust is derived primary from external sources through aerosol deposition and as a result of soil dusting, and less than household activities (Žibret, 2012; Žibret and Šajn, 2008a, 2008b, 2010; Šajn, 1999, 2000, 2001, 2002, 2003, 2005, 2006; Šajn et al., 2012). Usefulness of attic dust as a suitable long-term monitor for determination the status and content in air is proven by numerous studies (Alijagič and Šajn, 2001, 2006; Cizdziel et al., 2000; Ilacqua et al., 2003, Jeffrey et al., 2005; Gosar et al., 2006; Jemec Auflič and Šajn, 2007; Žibret, 2012; Žibret and Šajn, 2008a, 2008b, 2010; Šajn, 1999, 2000, 2001, 2002, 2003, 2005, 2006; Šajn et al., 2012).

Attic dust is commonly used as monitor for this process. These kind of monitoring programs have been performed as part of a large number of analytical studies for a longtime, but their application in recent decades has taken as wing. This is due to the fact that monitoring does not require the use of expensive technical equipment. Analytical results reflect the situation for a long period of time. The monitoring allows analysis of large areas and the types of monitors are chosen according to the analyst and the type of analysis.

Monitoring of air pollution has proved as a most useful technique for determining deposition of heavy metals and atmospheric pollution with it's in different geographical areas. The Republic of Macedonia does not deviate from the global framework of air pollution with heavy metals. The results obtained from previous studies on air pollution, suggest that the situation in the Republic of Macedonia is not favorable in terms of air pollution with heavy metals. The main emission sources appear to be power plants, mines, as well as flotation for lead, zinc and copper and metallurgical plants for the production of nickel, steel, lead, zinc and different ferroalloys (Barandovski et al., 2008; 2010, 2015; Stafilov et al., 2003, 2008a, 2008b, 2010, 2011, 2014).

Regarding the dominance of volcanic geological units and Pb-Zn mineral deposits, the region of Bregalnica river basin in the Republic of Macedonia, is a specific region to monitor the atmospheric distribution of various chemical elements (Stafilov et al., 2014; Balabanova et al., 2010, 2011, 2012; Bačeva et al., 2012; Barandovski et al., 2013, 2015). This region is characterized with several significant pollution sources of potentially toxic metals and other chemical elements in the environment: the copper mine and flotation "Bučim" near the town of Radoviš, the lead and zinc mines "Sasa" near the town of Makedonska Kamenica and "Zletovo" near the town of Probištip (Stafilov et al., 2014). The excavation of the copper minerals is carried out from an open ore pit, while in the lead-zinc mines the exploitation is underground, and the ore tailings are stored outdoors. The ore produced in the mines is processed in the flotation plants, and in the process of flotation of the relevant minerals, flotation tailings are separated and disposed on a dump site in the open.

Multivariate assessment and spatial hunting of dominant elements associations was conducted in the Bregalnica river basin and presents first attempt in this very lithologicaly unique area. The appearance of dominant Oligocene volcanism in the area of "Sasa" mine (Pb-Zn hydrothermal exploitation), Kratovo-Zletovo district and "Bučim" mine area (Cu-Au hydrothermal exploitation), it creates specific environmental conditions leading to natural and anthropogenic poly-metallic enrichments. Therefore, the main objective of this study was to assess the lithologenic and anthropogenic distribution of total of 69 elements along the Bregalnica river basin. Multivariate statistical approach was used for hunting the dominant geochemical associations. Moss species and attic dust samples were used for recording the lithological and anthropological impact on atmospheric deposition of potentially toxic and nontoxic elements. The dominant lithogenic and anthropogenic markers were extracted. The spatial distribution patterns of each geochemical marker were also generated.

2. Investigating Area

2.1. Geographic Characterization of the Investigated Area

The investigated area includes the basin of the Bregalnica river which is found in the area of the east planning region of the Republic of Macedonia. The investigated area covers ~200 km (W-E) x 200 km (S-N), that is, a total of ~4000 km^2, within the following geographic coordinates N: 41°27'-42°09' and E: 22°55'-23°01' (Figure 1). The region of the investigated area is geographically composed of several sub-regions. The area is characterized by two valleys – Maleshevska and Kočani valley. The Maleshevska valley represents the upper course of the Bregalnica river where the river source is also located, with average altitude of 700-1140 m. The valley is enclosed by the Maleshevski mountains on the east, by the Ograzhden mountain on south-southeast, and Plachkovica and Obozna on the west. The Malesevska valley is a specific geographical area, characterized by mountain, hilly, sloping and plain parts. The Kochanska valley includes the middle course of the Bregalnica river, between the mountains Plachkovica to the south and Osogovo to the north. The Maleshevski mountains are found to the east in relation to the valley, which is divided from them by the medium high mountains, Golak and Obozna. Morphologically, the valley is diverse, covering plain, hilly and mountain areas. The Kočani valley spreads on both sides of the Bregalnica river. The valley bottom that is, its plainest part occupies an area of 115 km^2 and represents an important agricultural cultivated fertile area. The lowest point above sea level in the area is located at the spot where the Zletovska river flows into the Bregalnica river (290 m).

The Bregalnica river is the central hydrographic factor in the eastern region of the R. Macedonia. The river's total length is 225 km, and the area of its basin amounts to 742 km^2. The more important tributaries are: from the right side – Zletovska, Kochanska and Orizarska river, and from the left side – Osojnica and Zrnovska river.

The source of the Bregalnica river is located in the forest area of the Maleshevski mountains near the Bulgarian border, east of Berovo, under the Chengino Kale peak, at altitude of about 1690 m. In its upper part it is characterized by a narrow canyon, the sides of which reach up to 360 m height above the river bed. The first erosive enlargement in this direction is the Ablanica area, at about 2 km from the town of Berovo, which is actually a spacious terrace positioned at about 2 m height from the water. The length of this enlargement is about 2 km, and the width 500-700 m. This whole area is intensively cultivated with cultures for the irrigation of which the water from the Bregalnica river is used. With the exception of this erosive enlargement, the narrow rocky part continues up to Berovo, with the river banks being stable and forested. Up to the village of Budinarci, the Bregalnica flows through the Berovo field. Because of the mild slope of the river bank, at this point the river flows rather slowly, enabling accumulation of gravel and sand, deposited in the river bed and around it. Here is also present the terrace of two to three meters, in which the river bed is positioned. From Budinarci all the way up to the point before the Delchevo field, the river bed is characterized with three rocky parts, among which there are two erosive enlargements. Starting right after Budinarci, and all the way to the village of Mitrashinci, the course of the Bregalnica river becomes curved, as a result of which at certain points real meanders and narrow rocky ravines are formed, the sides of which at certain points are vertical and with relative height of 60 to 80 m.

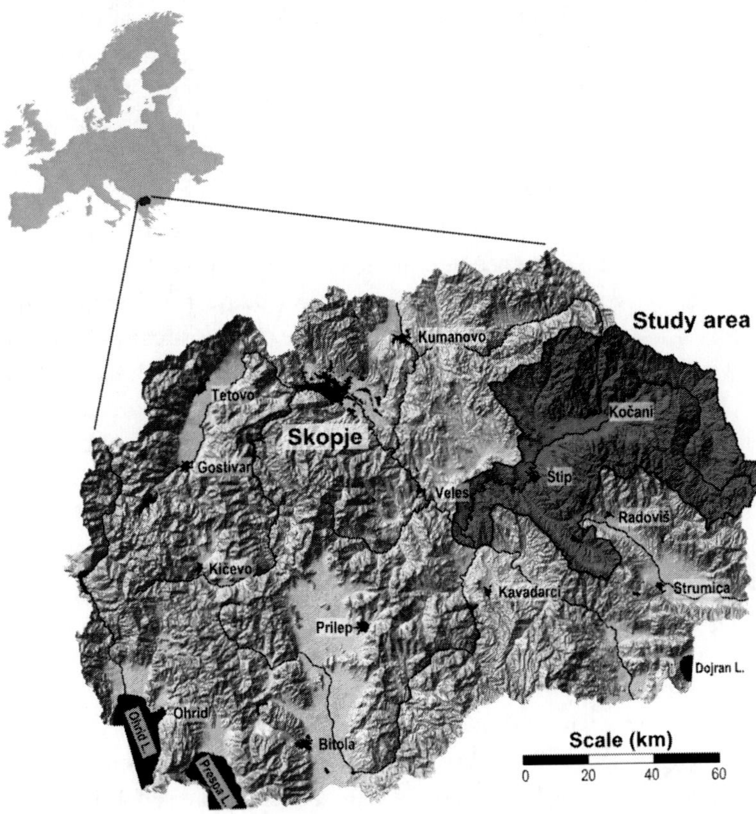

Figure 1. The investigated area on the territory of the Republic of Macedonia.

The second rocky part follows in the course of the river and spreads up to the village of Razlovci. It is also characterized by a narrow river bed, 10-20 m with curved watercourse and precipitous sides. The erosive enlargement at Razlovci, through which the Bregalnica flows, is about 5 km long and about 500 m with before the village, and after it about 800 m. The river does not have a single bed here, but the water is spread on the whole width, forming pools at certain points. Taking into account that along the river course a small and rather interesting canyon is formed, there are morphological and geological conditions for building of a dam that would enable accumulation in the Razlovci field. Entering into the Delchevo field, before the village of Trabotivishte, the Bregalnica river deposits huge quantities of material along the whole field, up to the mouth of the Ochepalska river, that is, to the entrance into the canyon along the river course. This deposit is mainly composed of bulky and fine gravel. At this point the course of the Bregalnica river is still small and does not have a single river bed, but instead several inlets are formed. In the Delchevo field the river waters are also extensively used for irrigation based on primitive dams and ditches, as for example, the ones near Trabotivishte and Delchevo. The water is almost completely used for irrigation of the cultivated low terrace (2 to 3 m) on the right river side (Mitrev et al., 2010).

From the mouth of the Ochipalska river, and all the way along the river course, there is a long and picturesque canyon that spreads up to the village of Istibanja, that is, to the entrance into the Kočani valley, about 35 km long. The canyon is poorly forested and hardly passable (today it is cut by a modern asphalt road), as its sides and banks are very steep, almost canyonlike, so that the valley is completely narrowed. Erosive enlargements are found only at the mouths of the larger rivers.

The Bregalnica flows into the Vardar river at Gradsko, at altitude of 137 m. Along the river course of the Bregalnica river two hydro accumulations were built: Ratevo (near Berovo) and Kalimanci (near Makedonska Kamenica), 80 m deep, 14 km long, 0,3 km wide, which accumulates 127 million m^3 water.

Taking into consideration the diverse morphological-geographical structure of the investigated area, the climate conditions also vary in the different sub-regions. Generally, the region is characterized with moderate continental climate. The altitude varies between 290 m in the plain of the Kočani valley up to 1932 m (Kadaica) on the Maleshevski mountains. The average annual temperature also varies in the different sub-regions. In the Kočani valley the average annual temperature is about 13°C, the warmest months in the year are July and August with average temperature of 25°C, and the coldest month is January with average temperature of about 1.2°C (Mitrev et al., 2010). In the region of Shtip, the average annual temperature is about 12.9°C, with high frequency of winds (out of 365 days in the year, there are air movements on 270 days). The Berovo valley can be singled out as the coldest region, which is under direct influence of the local mountain climate with an average annual temperature of 10°C (Environment statistics, 2013). Most common are the winds from western direction with frequency of 199‰ and speed of 2.7 m s^{-1} and the winds from eastern direction with frequency of 124 ‰ and speed 2.0 m s^{-1}.

The average annual precipitation amounts to about 500 mm with significant variations from year to year, as well as in the different sub-regions (Lazerevski, 1993). The precipitation is mostly related to and conditioned by the Mediterranean cyclones (Lazarevski, 1993). During the summer period, the region is most often found in the centre of the subtropical anticyclone, which causes warm and dry summers. From the central area of the region, as the driest area, the average annual precipitation increases in all directions, because of the increase either in the influence of the Mediterranean climate or the increase in altitude (Environment statistics, 2013).

The precipitation information refers to the annual average precipitation as measured by the meteorology stations in mm, information taken from the Hydrometeorological Administration. As regards the annual total number of sunny hours, there are about 6.0 hours annually in this area. In the region are distinguished about ten climatic-vegetation soil areas with considerably heterogeneous climate, soil and vegetation characteristics (Lazarevski, 1993).

Regarding the demographic structure in the region of the Bregalnica basin, only 4% of the populated areas belong to the category of urban areas, while 96% of the total populated areas are categorized as rural areas. In accordance with the data processed by the State Statistical Office (2013), the population density in the eastern region of R. Macedonia is <51 inhabitant on km^2. In accordance with the processed data, provided by Mitrev et al., (2010), in this region there is unequal distribution of urban and rural population. Considering land use, the region is considerably diverse. Along the whole course of the Bregalnica river dominate agricultural cultivated lands. Pastures are also considered as agricultural lands, and are represented in Figure 2, as open areas. About 30% belong to the forest regions, localized around the Maleshevska valley, represented by the Maleshevski Mountains, then around the Kočani valley, represented on one side by the Osogovo Mountains, and by Plackovica on the other side.

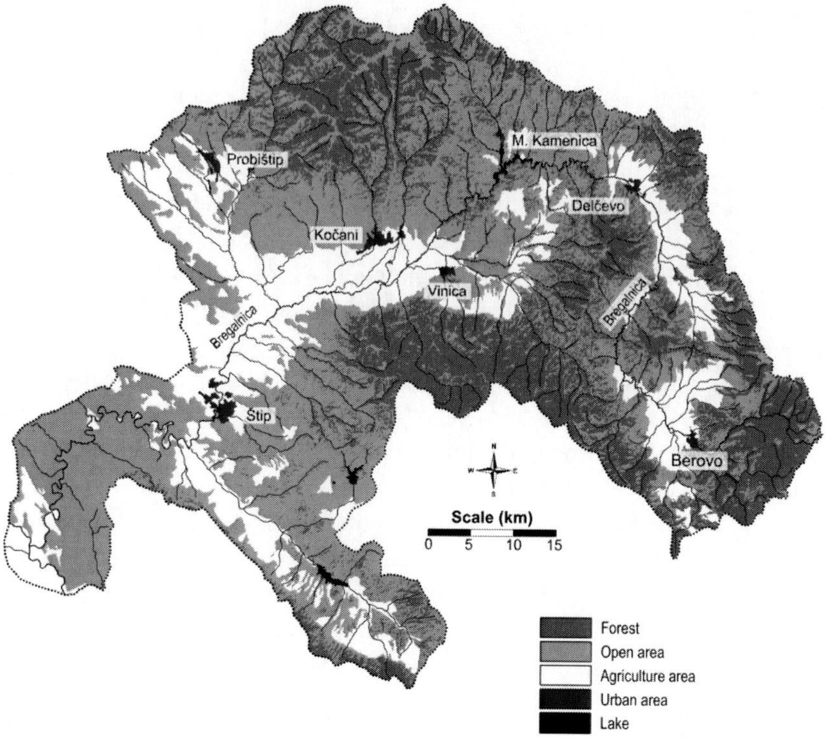

Figure 2. The region of the investigated area including the type of land use.

The hydrographic conditions show influence on the formation, characteristics and geography of soils through: surface flows, floods and irrigation. As a consequence of industrialization, urbanization and lack of treatment of wastewaters from the industry, mines and the city sewerage, the waters of this important hydrographic factor are exposed to a high level of pollution, from the aspect of introduction of higher contents of certain toxic metals. The pollution is most frequently particularly high at low river flows, as in the case of the Bregalnica river in its course through the Kočani valley.

2.2. Generalized Geology of the Investigated Area

The investigated area that covers the basin of the Bregalnica river lies on the two main tectonic units - the Serbian-Macedonian massive and the Vardar zone (Dumurdzanov et al., 2004). The polyphasal Neogene deformations through the insignificant movements associated with the volcanic activities had direct influence on the gradual formation of the reefs and the formation of deposits in the existing basins. From the middle Miocene to the end of the Pleistocene there were alternating periods of fast and slow landslide accompanied with variable sedimentation (deposition).

The Cenozoic volcanism represents a more recent extension in the Serbian-Macedonian massive and the Vardar zone. The oldest volcanic rocks occur in the areas of Bučim, Damjan, the Borov Dol district and in the zone of Toranica, Sasa, Delchevo and Pehchevo (Dumurdzanov et al., 2004). These older volcanic rocks were formed in the mid Miocene from sedimentary rocks that represent the upper age limit of the rocks. The origin of these oldest volcanic rocks is related to the Oligocene – the early Miocene period.

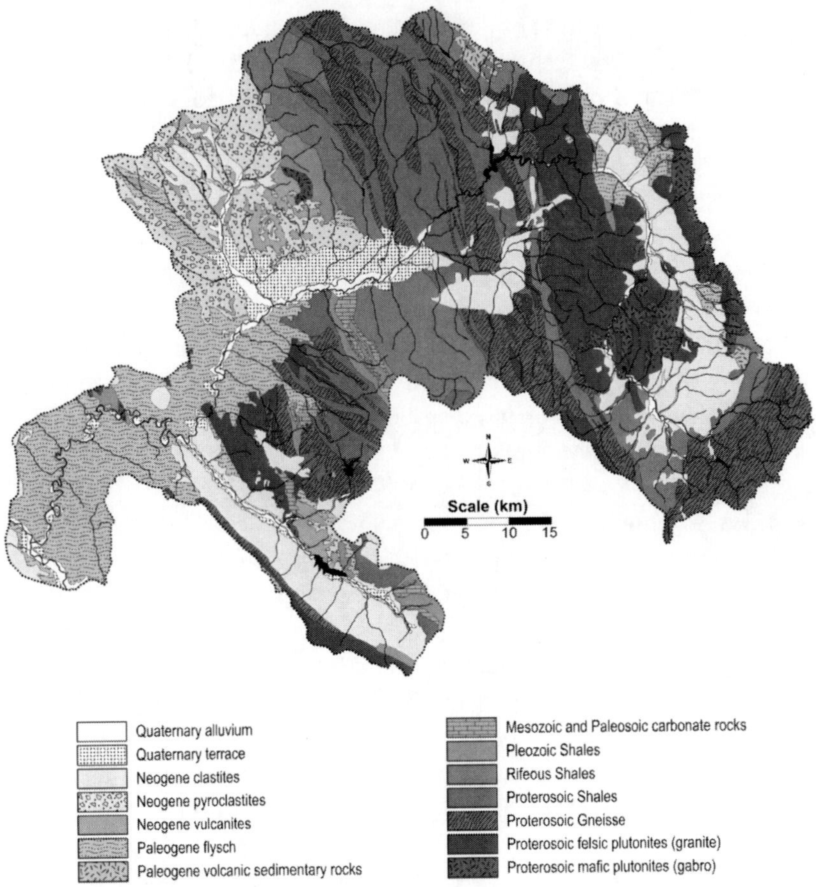

Figure 3. Generalized geology of the investigated area.

As volcanic rocks are categorized the following: andesite, latite, quartz-latite, dacite. Volcanism appears sequentially and in several phases forming sub-volcanic areas. On the other hand, the pyroclastites are most frequently found in the Kratovo-Zletovo volcanic area, where the dacites and andesites are the oldest formations. Generalized geology of the area is produced based on the data provided by Rakićević et al., (1968) and presented in Figure 3.

3. Materials and Methods

3.1. Dataset Building Sampling Strategy and Analytical Methods

The monitoring activities implemented included determining of the contents of a total of 69 elements: Ag, As, Al, Au, B, Ba, Be, Bi, Br, Ca, Cd, Ce, Co, Cr, Cs, Cu, Dy, Er, Eu, Fe, Ga, Gd, Ge, Hf, Hg, Ho, I, In, Ir, K, La, Li, Lu, Mg, Mn, Mo, Na, Nb, Nd, Ni, Os, P, Pb, Pd, Pr, Pt, Rb, Re, Rh, Ru, Sb, Sc, Se, Sm, Sn, Sr, Ta, Tb, Te, Ti, Th, Tl, Tm, V, W, Y, Yb, Zn и Zr. The content of these elements was determined in moss samples and attic dust samples.

For each location where from samples were taken the location characteristics were recorded (geographic coordinates and altitude) using a global positioning system. This positioning, that is, specifying the position of the locations where from samples for investigation were taken, is required for the construction of maps for the distribution and deposition of the analyzed elements in the investigated area.

3.1.1. Moss Samples

The carpet-forming moss species *Pleurozium schreberi* and *Hylocomium splendens* are usually preferred for moss trace elements deposition monitoring (Fernández et al., 2015). But these species are commonly found only in some parts of the country at an elevation above 1000 m (Cekova, 2005). However, the altitude of the investigated area varies between 290 to 1932 m. Therefore, the samples of the pleurocarpous moss species *Homalotecium lutescens* and *Hypnum cupressiforme* were also collected in the investigated area in the period of August-September 2012. Researchers setting up large-scale survey often face the problem that the location of the predicted sampling spot becomes subordinate to the presence/absence of the selected species (Fernández et al., 2015). This problem can be overcome by using more than one moss species within the same survey; however, it is clear that the concentrations of elements may vary considerably between species thus precluding comparison of the results obtained (Market, 2007). Carballeira et al., (2008) suggested that when the regression line slope and elevation do not differ significantly from the line of slope 1 (covariance analysis), the species could be combined. Considering this four species were used for conducting the monitoring for smaller range scale deposition.

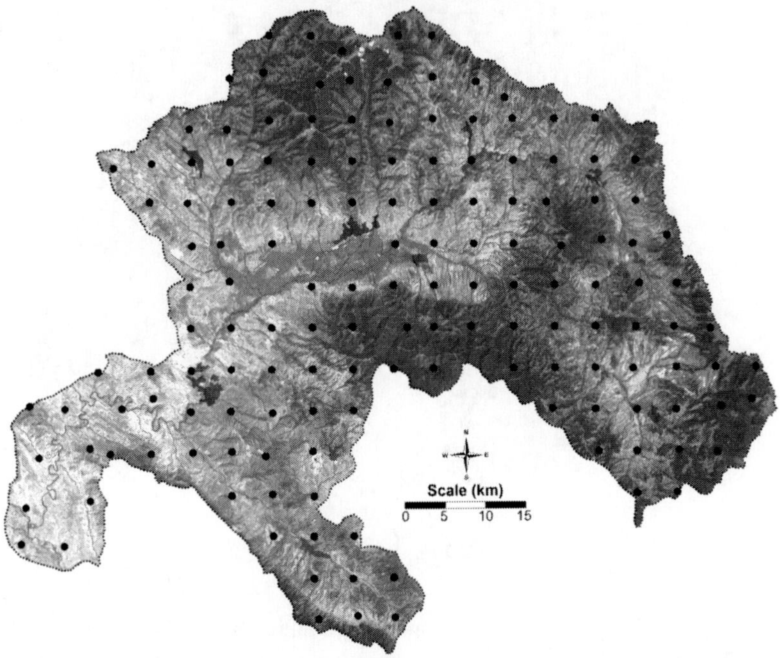

Figure 4. Digital elevation model and location of sampling points for moss samples.

Depending on the conditions and the accessibility of the locations the specie that is available and typical for the region was collected. Random samples (in the very close vicinity of the pollution source) and samples according to sampling network (5 x 5 km) were collected from total 149 sample location, as presented in Figure 4. Detailed description of the collection of samples (according officially accepted techniques) is given by Fernández et al., (2015).

3.1.2. Attic Dust Samples

Attic dust samples were collected from the attic of total 84 houses built between 1920 and 1970. The collection of attic dust samples was performed according to the adopted protocol form Šajn (2005). Dust samples were collected with plastic brush in polyethylene bags. At each location for sampling attic dust, topsoil (0-5 cm) samples from the house yards were collected (Figure 5).

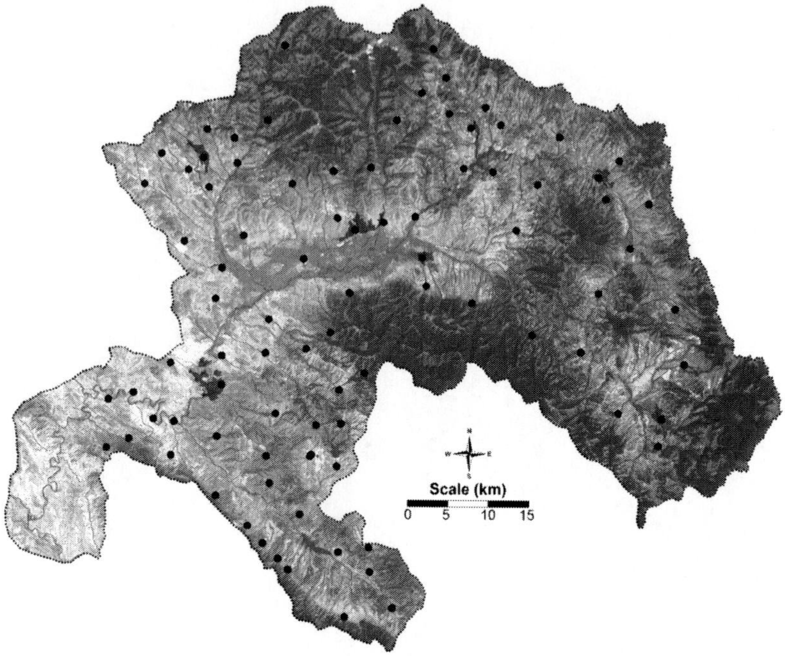

Figure 5. Digital elevation model and location of sampling points for attic dust and top soil samples.

3.2. Analytical Reagents

All used reagents and standards are with the following analytical purity level: nitric acid, HNO_3, 69%, ultra pure (Merck, Germany), hydrogen peroxide, H_2O_2, 30 % p.a. (Merck, Germany), concentrated hydrochloric acid, HCl 37%, p.a. (Merck, Germany), concentrated hydrofluoric acid, HF, p.a. (Merck, Germany) and concentrated perchloric acid, $HClO_4$, p.a. (Alkaloid Skopje). Redistilled water was used for the preparation of all solutions. The standard solutions for the elements examined were prepared by dissolution of the basic multi element standard solution with concentration of 1000 mg L^{-1} (11355-ICP, Multi Element Standard). A series of standard solutions were prepared in a linear range in several concentration areas as, *series 1*: 1, 2, 3, 5, 10 μg L^{-1} and *series 2*: 10, 20, 30, 50, 100 μg L^{-1} for elements in traces (determined using MS-ICP), for the remaining elements; *series 3*: 0.1; 0.2; 0.3; 0.5; 1 mg L^{-1}; *series 4*: 1, 2, 3, 5, 10 mg L^{-1} and *series 5*: 10, 20, 30, 50, 100 mg L^{-1} for the elements in macro contents.

3.3. Sample Preparation

For digestion of moss samples, the microwave digestion system (CEM, model Mars) was applied. Precisely measured mass (0.5000 g) of moss samples, than 5 mL concentrated HNO_3 (trace pure), and 2 mL H_2O_2 (30%, m/V) were added. The Teflon vessels were carefully closed and the microwave digestion method was applied. Digestion method was performed in to two steps for total dissolving of moss tissue as previously given by Balabanova et al., (2010) and Bačeva et al., (2012). After the digestion method was finished, digests were quantitatively transferred into 25 ml volumetric flaks. Thus way prepared digests from moss tissue were analysed for the total elements contents.

For digestion of attic dust and soil samples, open wet digestion with mixture of acids was applied (ISO 14869-1:2001). Precisely measured mass of dust samples (0.5 g) was placed in teflon vessels and 5 mL concentrated nitric acid, HNO_3 was adding, until the brown vapours came out from the vessels. Nitric acid is very suitable oxidant for digestion of environmental samples. For total digestion of inorganic components was added 5-10 mL hydrofluoric acid. When the digest became clear solution, 2 mL of $HClO_4$ was added. Perchloric acid was used for total digestion of organic matter. After 15 min. cooling the vessels, 2 mL of HCl and 5 mL of H_2O were added for total dissolve of metal ions. Finally the vessels were cooled and digests quantitatively transferred to 50 mL calibrated flasks. In this way the digested soil and sediment samples were prepared for determining the contents of the different elements using atomic emission and mass spectrometry.

3.4. Instruments Used for Determining the Content of Elements in Various Samples

The analysis of elements content in digested samples was conducted with application of atomic emission spectrometry with inductively coupled plasma

(AES-ICP) and mass spectrometry with inductively coupled plasma (MS-ICP). For each element analyzed, previous optimization of the instrumental conditions was performed. In all samples, the contents of a total of 23 elements were analyzed: Ag, Al, As, B, Ba, Ca, Cd, Co, Cr, Cu, Ga, Fe, K, Li, Mg, Mn, Mo, Na, Ni, Pb, Sr, V and Zn by applying AES-ICP. By applying mass spectrometry with inductively coupled plasma a total of 69 isotopes were analyzed, which are the following: Ag, As, Al, Au, B, Ba, Be, Bi, Br, Ca, Cd, Ce, Co, Cr, Cs, Cu, Dy, Er, Eu, Fe, Ga, Gd, Ge, Hf, Hg, Ho, I, In, Ir, K, La, Li, Lu, Mg, Mn, Mo, Na, Nb, Nd, Ni, Os, P, Pb, Pd, Pr, Pt, Rb, Re, Rh, Ru, Sb, Sc, Se, Sm, Sn, Sr, Ta, Tb, Te, Ti, Th, Tl, Tm, V, W, Y, Yb, Zn, Zr.

The instrumental and operating conditions for each of the above mentioned techniques are given in Table 1.

The limits of detection (LOD) were based on the usual definition as the concentration of the analytic yielding a signal equivalent to three times the standard deviation of the blank signal, using 10 measurements of the blank for this calculation. The calculated values for the detection limits are given in Table 2.

SCIEX Perkin Elmer Elan DRC II (Canada) inductively coupled plasma mass spectrometer (ICP-MS, with quadrupole as single detector) was used for measurement of the concentration trace elements, while for the major elements content determination atomic emission with inductively coupled plasma (ICP-AES, Varian 715ES) was applied.

Certified reference materials NIST 2709 (for attic dust samples), M2 and M3 (for moss samples) were used to check method accuracy. For all considered elements, the difference between measured and certified values was within 15%. The theoretical limit for ICP-MS methods are in ppt (ng/L) range for the majority of the elements. Matrix effects above 1 ppb (μg/L) threshold while using Total Quant were not observed during our study. For some elements, values between these two levels were further investigated using more complex quantitative methods. Spiked intra-laboratory samples were analyzed at a combined frequency of 20% of the samples. The recovery for all of the analyzed elements ranges from 76.8% for Tl to 119% for Sb (for ICP-MS measurements) and from 87.5% for Na to 112% for P (for ICP-AES measurements).

Table 1. Instrumental conditions for ICP-AES and ICP-MS

<table>
<tr><td colspan="2">RF Generator</td><td>ICP-AES</td><td>ICP-MS</td></tr>
<tr><td colspan="2">Power output of RF generator</td><td colspan="2">1500 W</td></tr>
<tr><td colspan="2">Power output stability</td><td colspan="2">Better than 0.1%</td></tr>
<tr><td colspan="2">ICP Ar flow gas rate</td><td colspan="2">15 L min^{-1}</td></tr>
<tr><td colspan="2">Plasma parameters</td><td></td><td></td></tr>
<tr><td colspan="2">Nebulizer</td><td>V- groove</td><td>Micromist</td></tr>
<tr><td colspan="2">Spray Chamber</td><td colspan="2">Double-pass cyclone</td></tr>
<tr><td colspan="2">Peristaltic pump</td><td colspan="2">0-50 rpm</td></tr>
<tr><td colspan="2">Cones</td><td>/</td><td>Platinum</td></tr>
<tr><td colspan="2">Plasma configuration</td><td>Radially viewed</td><td>Axially viewed</td></tr>
<tr><td colspan="2">Spectrometer</td><td>Echelle optical design</td><td>Quadrupole</td></tr>
<tr><td colspan="2">Polychromator</td><td>400 mm focal length</td><td>/</td></tr>
<tr><td colspan="2">Polychromator purge</td><td>0.5 L min^{-1}</td><td>/</td></tr>
<tr><td colspan="2">Total voltage /V</td><td>/</td><td>0.1</td></tr>
<tr><td colspan="2">Integration measurement time /ms</td><td>/</td><td>0.1</td></tr>
<tr><td colspan="2">Measurement at one point (isotope)/s</td><td>/</td><td>300</td></tr>
<tr><td colspan="2">Repetitions measurement</td><td colspan="2">3 per point</td></tr>
<tr><td colspan="4">Conditions for program</td></tr>
<tr><td colspan="2">ICP-AES measurements</td><td colspan="2">ICP-MS measurements</td></tr>
<tr><td>Element</td><td>Wavelength, nm</td><td colspan="2">Isotopes</td></tr>
<tr><td>Al</td><td>396.152</td><td colspan="2" rowspan="7">^{107}Ag, ^{75}As, ^{27}Al, ^{197}Au, ^{11}B, ^{137}Ba, ^{9}Be, ^{209}Bi, ^{79}Br, ^{114}Cd, ^{140}Ce, ^{59}Co, ^{53}Cr, ^{133}Cs, ^{63}Cu, ^{163}Dy, ^{166}Er, ^{153}Eu, $^{56/57}$Fe, ^{69}Ga, ^{157}Gd, ^{72}Ge, ^{178}Hf, $^{201/202}$Hg, ^{165}Ho, ^{127}I, ^{115}In, ^{193}Ir, ^{139}La, ^{7}Li, ^{175}Lu, ^{55}Mn, ^{95}Mo, ^{93}Nb, ^{146}Nd, ^{60}Ni, ^{189}Os, $^{206/207/208}$Pb, ^{105}Pd, ^{141}Pr, ^{195}Pt, ^{85}Rb, ^{185}Re, ^{103}Rh, ^{101}Ru, ^{121}Sb, ^{45}Sc, ^{77}Se, ^{147}Sm, ^{120}Sn, ^{88}Sr, ^{181}Ta, ^{159}Tb, ^{125}Te, ^{47}Ti, ^{232}Th, ^{205}Tl, ^{169}Tm, ^{51}V, ^{182}W, ^{89}Y, ^{172}Yb, ^{66}Zn, ^{90}Zr</td></tr>
<tr><td>Ca</td><td>370.602</td></tr>
<tr><td>Fe</td><td>238.204</td></tr>
<tr><td>Mg</td><td>280.270</td></tr>
<tr><td>K</td><td>766.491</td></tr>
<tr><td>Na</td><td>589.592</td></tr>
<tr><td>P</td><td>213.618</td></tr>
</table>

Table 2. Calculated instruments lower detection limits for the analyzed elements

Element	Limit of detection, μg L^{-1}
Ag, Au, Cs, Dy, Eu, Hf, Ho, In, Ir, Li, Lu, Os, Pd, Pt, Re, Ru, Ta, Tb, Te, Ti, Tl, Tm, W	0.001
Rh	0.002
As, Al, B, Ba, Be, Bi, Br, Cd, Co, Cr, Cu, Eu, Fe, Ga, Hg, I, K, La, Mg. Mn, Mo, Na, Nb, Nd, Ni, P, Pr, Rb, Sb, Sc, Se, Sm, Sn, Sr, Th, V, Y, Yb, Zn, Zr	0.010
Ce	0.050
Pb	0.100

3.5. Data Processing

The obtained values for the contents of the investigated elements were statistically processed using basic descriptive statistics. The application of bivariate statistic was used for data check of elements contents correlations. For that issue the linear coefficient of correlation was used. Two-dimensional scatter-plots were used to visualize relations between two data sets. Individual data points were represented by point markers in two-dimensional space, where axes represent the variables. Multivariate statistic method (cluster and R-mode factor analyses) was used to reveal the associations of the chemical elements. The factor analysis was performed on variables standardized to zero mean and unit standard deviation (Šajn, 1999, 2000, 2001, 2002, 2003, 2005, 2006). As a measure of similarity between variables, the product-moment correlation coefficient (*r*) was applied. There are various rotational strategies that have been proposed (Žibret and Šajn 2010; Šajn, 1999, 2000, 2001, 2002, 2003, 2005, 2006). The goal of all of these strategies is to obtain a clear pattern of loadings, factors that are somehow clearly marked by high loadings for some variables and low loadings for others. The elements with low communalities were excluded because of their lack of significant associations. In this study, the *varimax* method was used for orthogonal rotation (Sakata et al., 2003; Kaymaz, 2005). As before, we want to find a rotation that maximizes the variance on the new axes; put another way, we want to obtain a pattern of loadings on each factor that is as diverse as possible, lending itself to easier interpretation Statistical softer *Statistica 11.0*, was used for data processing of obtained values for elements contents.

4. Results and Disscusion

4.1. Monitoring Distribution of Elements in Moss Samples

4.1.1. Summary Results of the Data Set

The assessments for the elements content distribution in the investigated area was characterized using values obtained from basic descriptive statistics (Table 3). The distribution of macroelements Al, Ca, Fe, K, Mg and P was in the range of worthwhile-ness typical of this kind of samples. High matrix content originated from plant tissues can sometimes enriches the actual values from atmospheric distribution of these elements. The determined median values for Al, Ca, Fe, K, Mg and P are: 0.27%; 0.94%; 0.26%; 0.22%; 0.17% and 0.14%, respectively (Table 3). Compared to data published from Barandovski et al., (2015) and Harmens et al., (2015) for the whole of R. Macedonia and the territory of Europe no significant variation of these elements occurs. The distribution of the rest of the elements was due to certain natural phenomena (geology of the area, winds, rain, anthropogenic activities, etc.). Silver content was determined in the range of 0.005-1.1 mg kg^{-1}, with a median of 0.038 mg kg^{-1} (Table 2). Native silver is sometimes associated with sulfide ores as a result of their chemical reduction. In the area of the copper sulfides hydrothermal exploitation occur in area of Proterosoic shales, the obtained median value for Ag was 0.051 mg kg^{-1}. Maximum median for Ag, accordingly to dominant geological units was obtained in moss samples from Quaternary terraces (Table 3). Very similar to Ag, distributes Cu in the investigated area, also enriched in the Quaternary terraces (median value of 6.7 mg kg^{-1} (Table 3). These metals, collectively known as the "coinage metals," have very similar geochemistry associated with their organometallic chemistry (Greenwood and Earnshaw, 2005).

For As, Sb and Bi no significant enrichments were obtained for the atmospheric deposition in moss. None of these three elements do not participate significantly in the composition of the Earth's crust, although several minerals containing as its major constituents. Median value for arsenic content (0.47 mg kg^{-1}) compared to data from the European report published from Harmens et al., (2015) was lower than the median value obtained for whole of R. Macedonia. Arsenic content compared to other European countries, such as Norway (0.13 mg kg^{-1}), France (0.18 mg kg^{-1}) or Finland (0.11 mg kg^{-1}) was significantly enriched. Dumurdzanov et al., (2004) explained that natural enrichment of arsenic may occur in areas where the Neogene vulcanite's are dominant geological units. Bismuth content ranges from 0.005-1.9 mg kg^{-1}, with predominant distribution in areas with Neogene volcanism (median value 0.054 mg kg^{-1}). For antimony was obtained median value of 0.019 mg kg^{-1}. Investigation conducted on the whole territory of the Republic of Macedonia by Barandovski et al., (2015) showed that Sb enrichments are with anthropogenic origin, occurring in areas with Pb-Zn mineral hydrothermal exploitation.

Boron is the element with very specific chemistry in the environment. The content of B in the moss samples ranges from 1.8-59 mg kg^{-1} and a median of 13 mg kg^{-1}. The atmospheric distribution of boron in the investigated area was intensive on Paleogene flisch (median 20 mg kg^{-1}). Very similar to boron distributes the alkaline earth metals, such as Be, Ba and Sr. Barium in moss samples ranges from 14-350 mg kg^{-1}. Beryllium was dominantly enriched on the Quaternary terraces and in the areas with dominant Ng-volcanism (0.13 mg kg^{-1}), while the whole investigated area was characterized with Be deposition in range of 30-350 μg kg^{-1}, with median value of 96 μg kg^{-1} (Table 3). To this group of elements can be incorporated also lithium, cesium and rubidium. These elements showed very low variation in the deposition accordingly to the different geologic units. These alkaline metals have very good stability in marking the lithogenic distribution. The investigated area was characterized with medians for Li, Cs and Rb as follow: 2.6, 0.27 and 5.4 mg kg^{-1}, respectively (Table 3).

Gallium and thallium usually are correlated with the occurrence of vulcanite's; it has also been demonstrated in present investigation. Depositions of these elements relays dominantly in area of Neogene pyroclastites and vulcanite (Kratovo-Zletovo district). Characterization of the study area accordingly to dominant geological units, extracted Tl with 61 mg kg^{-1} in area of Neogene pyroclastites. River terraces were also enriched with Ga and Tl (1.3 mg kg^{-1} and 59 μg kg^{-1}, respectively). The halogens, bromine and iodine deposited with median values of 1.1 mg kg^{-1} and 91 μg kg^{-1}, respectively. These elements shows very high reactivity in environment and usually their deposition is correlated to geochemical occurrence as potassium and sodium salts. River terraces are mostly enriched with their contents, as in the present case. However, no significantly variation occurs in accordingly dominant geological units.

Table 3. Descriptive statistic parameters for elements content in moss samples

	Unit	Mean	Med	Min	Max	P_{25}	P_{75}	*S*	*CV*	*A*	*E*
Al	%	0.27	0.27	0.080	1.5	0.18	0.42	0.018	67	0.19	-0.35
Ag	µg/kg	37	38	5.0	1100	24	55	12	200	-0.14	1.51
As	mg/kg	0.48	0.47	0.05	4.4	0.29	0.86	0.058	100	-0.02	0.46
B	mg/kg	13	13	1.8	59	8.1	19	0.82	67	0.03	-0.20
Ba	mg/kg	48	46	14	350	37	66	3.4	71	0.07	1.16
Be	µg/kg	95	96	30	530	64	150	6.2	65	-0.04	-0.43
Bi	µg/kg	32	35	5.0	1900	20	51	14	270	-0.16	1.68
Br	mg/kg	1.3	1.1	0.14	15	0.72	2.4	0.20	110	0.02	-0.38
Ca	%	0.95	0.94	0.29	2.1	0.77	1.1	0.026	33	0.41	0.74
Cd	µg/kg	82	80	32	2200	60	120	25	190	0.09	-0.14
Ce	mg/kg	3.6	3.6	0.93	24	2.3	5.9	0.29	75	0.05	-0.49
Co	mg/kg	1.0	1.1	0.31	4.0	0.65	1.6	0.067	64	0.03	-0.70
Cr	mg/kg	4.2	4.3	0.85	32	2.5	6.4	0.44	91	0.04	-0.08
Cs	mg/kg	0.27	0.27	0.057	5.1	0.16	0.49	0.052	130	0.03	-0.49
Cu	mg/kg	4.5	4.4	2.2	240	3.7	5.8	1.8	270	0.13	0.21
Dy	mg/kg	0.37	0.38	0.079	1.7	0.23	0.60	0.027	70	-0.02	-0.38
Er	mg/kg	0.18	0.19	0.041	0.87	0.12	0.30	0.013	70	-0.03	-0.31
Eu	µg/kg	84	85	22	390	50	130	5.9	68	-0.04	-0.60
Fe	%	0.25	0.26	0.077	1.2	0.15	0.39	0.016	65	0.18	-0.56
Ga	mg/kg	0.90	0.93	0.28	5.0	0.62	1.3	0.060	66	0.01	-0.36
Gd	mg/kg	0.48	0.50	0.11	2.6	0.29	0.76	0.036	72	-0.01	-0.47
Ge	µg/kg	25	25	11	91	18	34	1.1	49	-0.09	-0.40
Hf	µg/kg	37	36	5.0	340	27	49	3.5	90	-0.14	2.70
Ho	µg/kg	73	74	16	340	47	120	5.2	69	-0.05	-0.32
I	µg/kg	92	91	38	260	75	130	3.1	37	-0.08	-0.38
K	%	0.22	0.22	0.067	0.66	0.18	0.25	0.006	34	0.45	3.45
La	mg/kg	1.7	1.7	0.42	14	1.0	2.8	0.16	87	0.03	-0.42

Table 3. (Continued)

	Unit	Mean	Med	Min	Max	P_{25}	P_{75}	*S*	*CV*	*A*	*E*
Li	mg/kg	2.5	2.6	0.70	20	1.7	3.8	0.20	75	0.03	-0.22
Lu	µg/kg	25	26	5.0	120	16	41	1.7	70	-0.08	-0.14
Mg	%	0.16	0.17	0.058	0.42	0.13	0.20	0.005	37	0.30	0.41
Mn	mg/kg	150	150	32	810	100	220	9.4	64	0.12	0.14
Mo	mg/kg	0.22	0.22	0.063	4.2	0.16	0.34	0.038	140	-0.07	0.53
Na	mg/kg	58	59	28	2300	47	76	15	230	0.10	0.60
Nd	mg/kg	1.8	1.8	0.44	10	1.0	2.9	0.14	72	0.03	-0.61
Ni	mg/kg	4.0	4.1	1.6	45	3.0	5.6	0.51	110	0.06	0.05
P	%	0.14	0.14	0.041	0.29	0.11	0.17	0.004	34	0.29	0.25
Pb	mg/kg	5.4	4.9	2.0	1400	3.6	8.5	17	360	0.43	-0.35
Rb	mg/kg	5.5	5.4	1.9	47	3.7	8.3	0.40	73	0.06	-0.25
Sb	µg/kg	18	19	5.0	96	13	27	1.3	69	-0.08	0.12
Sc	mg/kg	1.2	1.2	0.55	6.5	0.93	1.7	0.066	56	0.03	-0.22
Sm	mg/kg	0.37	0.37	0.050	2.0	0.21	0.60	0.027	71	-0.02	-0.12
Sn	mg/kg	0.20	0.19	0.025	1.8	0.16	0.25	0.018	85	-0.11	3.82
Sr	mg/kg	27	26	9.4	110	20	41	1.6	59	0.10	-0.29
Tb	µg/kg	66	68	15	310	40	100	4.8	70	-0.05	-0.45
Ti	mg/kg	120	130	32	570	74	190	8.2	67	0.10	-0.42
Tl	µg/kg	39	36	5.0	280	25	65	4.0	90	-0.07	0.33
Tm	µg/kg	28	28	5.0	130	18	44	2.0	70	-0.07	0.06
V	mg/kg	5.9	6.0	1.8	32	3.6	8.8	0.43	71	0.06	-0.52
W	µg/kg	62	63	5.0	1300	24	140	17	150	-0.03	-0.26
Y	mg/kg	1.5	1.5	0.35	7.6	0.97	2.5	0.11	71	0.01	-0.38
Yb	mg/kg	0.16	0.17	0.033	0.77	0.10	0.26	0.01	69	-0.04	-0.29
Zn	mg/kg	16	16	6.6	240	13	20	2.6	140	0.08	0.99
Zr	mg/kg	1.1	1.0	0.037	13	0.80	1.5	0.11	99	0.05	5.36

Med – median; Min – minimum; Max – maximum; S – standard deviation; CV – coefficient of variation; A – skewness; E – kurtosis.

Chromium and nickel shows a great similarity in distribution in the range of 0.85 to 32 and 1.6-45 mg kg^{-1}, respectively (Table 3). Their distribution is dominant on Quaternary terraces, probably due to the accumulation of soil dust from soil formed in Paleogene flysch.

The distribution of the LREEs (La-Ce-Pr-Nd-Sm-Gd) is predominantly related to the Quaternary terraces and the Paleogene flysch. The anthropogenic activities that occur in the Pb-Zn surroundings and the Cu mineralization in the investigated area significantly influence the lithological distribution of this geochemical association of elements. On the other hand, the long-term deposition of the HREEs (Eu-Tb-Dy-Ho-Er-Tm-Yb-Lu) presents a typical geochemical marker of the area of the Bregalnica river basin (Balabanova et al., 2015).

The latest results reported from the UNECE ICP Vegetation – Heavy Metals in European Mosses (Barandovski et al., 2015; Harmens et al., 2015) indicate on one of the most health risk area with higher values of Cd, Pb and Zn. That area has been found in the eastern parts of Macedonia, due to the operation conducted in mines for Pb-Zn hydrothermal exploitation (Sasa mine and Zletovo mine). Lead and Zn are specifically distributed around Kamenička and Zletovska River, areas with dominant occurrence of Neogene and Paleogene volcanism. The intensive deposition of these elements includes maximum content in moss of 1400 mg kg^{-1} for Pb and 240 mg kg^{-1} for Zn (Table 3). Cadmium content vary in range of 0.032-2.2 mg kg^{-1}, compared to the whole territory of the Republic Macedonia (median 0.22 mg kg^{-1}), this element has a reduced content distribution (Barandovski et al., 2013; Harmens et al., 2015).

4.1.2. Multivariate Assessing for Dominant Geochemical Markers

Multivariate statistical *hunting* was applied for reducing the numerous elements distribution and identifying the dominant geochemical association. Extraction was simplified to 7 factor association, marked as: F1: Al-Be-Co-Fe-Ga-Ge-Li-Mg-Sc-Ti-V-Y-(La-Gd)-(Eu-Lu); F2: Ba-Bi-Cd-Pb-Sb-Zn; F3: As-Cs-Rb-Tl; F4: Ca-Sr-B; F5: Hf-Zr; F6: Cr-Cu-Ni; F7: Br-I, with a total variability for dominant loadings of 84%. Table 4 summaries factor loadings for dominant elements associations. Cluster analysis was also applied, in order to determine the functional dependence of internal associations (Figure 6).

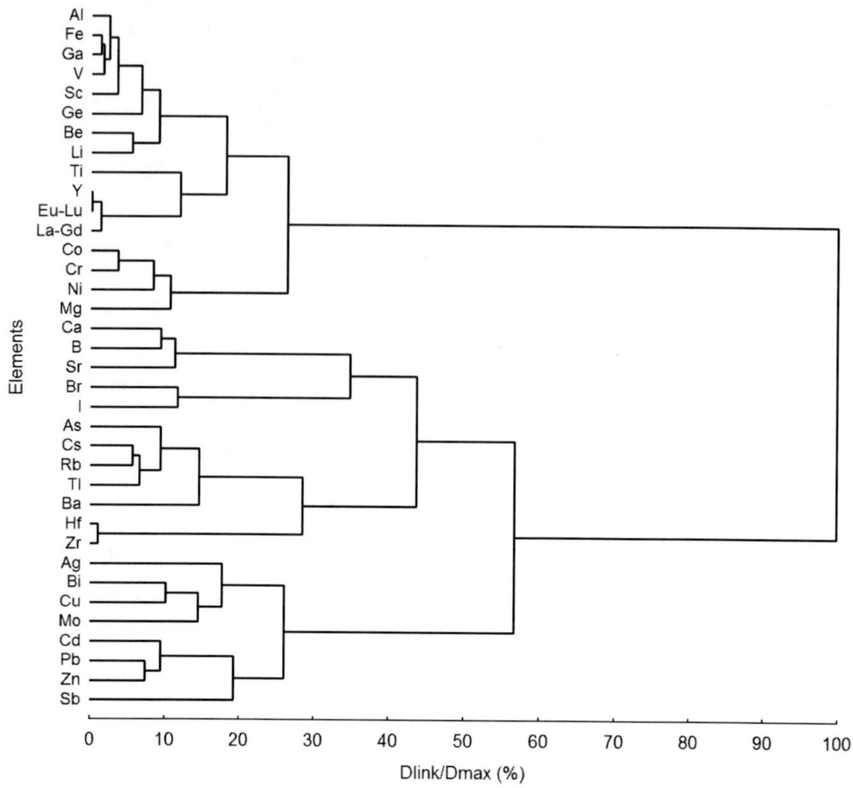

Figure 6. Dendogram for dominant clusters.

4.1.3. Areal Distribution Mapping of the Factors

For the spatial distribution of the factor scores, universal kriging with the linear variogram interpolation method was applied for the construction of maps showing the spatial distribution of factor scores (F1-F7).

Factor 1: [Al-Be-Co-Fe-Ga-Ge-Li-Mg-Sc-Ti-V-Y-(La-Gd)-(Eu-Lu)] can possibly be explained by elements associated with mineral particles, mainly windblown dust, and contributes to 30 % of the variance. The highest values of the factor loading were obtained for rare earth elements, (0.95 and 0.90). The distribution of these elements represents a specific lithological phenomenon, due to the strong dominance rare earth elements (Balabanova et al., 2015).

Table 4. Matrix of dominant factor loading

Element	F1	F2	F3	F4	F5	F6	F7	Communality
Al	0.76	0.14	0.38	0.19	0.24	0.23	0.07	89.0
Be	0.68	0.12	0.50	0.23	0.27	0.06	0.11	87.0
Co	0.73	0.10	0.19	0.20	0.11	0.50	0.19	92.0
Fe	0.82	0.11	0.30	0.14	0.17	0.34	0.13	95.9
Ga	0.81	0.10	0.37	0.14	0.27	0.22	0.15	96.9
Ge	0.70	0.05	0.14	0.28	0.43	0.23	0.13	84.6
Li	0.52	0.01	0.48	0.40	0.19	0.34	0.11	82.0
Mg	0.49	0.11	0.23	0.57	0.07	0.36	0.11	78.4
Sc	0.79	0.10	0.22	0.15	0.25	0.30	0.05	85.5
Ti	0.78	0.21	0.03	0.01	0.14	0.25	0.07	75.2
V	0.74	0.12	0.31	0.20	0.28	0.35	0.15	92.9
Y	0.95	0.07	0.14	0.10	0.02	0.04	0.07	94.5
La-Gd	0.90	0.10	0.22	0.13	0.09	0.01	0.12	91.0
Eu-Lu	0.95	0.09	0.15	0.08	0.05	0.03	0.06	95.4
Ba	0.35	0.52	0.23	0.22	0.37	0.26	0.19	73.5
Bi	0.37	0.55	0.31	0.01	0.07	0.31	0.25	70.1
Cd	0.04	0.88	0.10	0.06	0.06	0.09	0.08	81.2
Pb	0.08	0.81	0.32	0.06	0.16	0.10	0.10	80.9
Sb	0.51	0.63	0.09	0.10	0.15	0.17	0.09	73.7
Zn	0.25	0.82	0.15	0.05	0.08	0.22	0.00	81.8
As	0.40	0.34	0.55	0.24	0.09	0.25	0.21	75.3
Cs	0.33	0.21	0.77	0.27	0.14	0.03	0.14	86.4
Rb	0.47	0.24	0.66	0.07	0.30	0.07	0.10	82.2
Tl	0.41	0.31	0.60	0.12	0.35	0.03	0.26	82.8
Ca	0.21	0.06	0.01	0.85	0.01	0.18	0.13	81.7
B	0.16	0.02	0.22	0.81	0.01	0.09	0.04	73.7
Sr	0.02	0.18	0.13	0.78	0.38	0.07	0.09	82.0
Hf	0.34	0.09	0.19	0.13	0.87	0.11	0.03	94.9
Zr	0.25	0.15	0.21	0.11	0.88	0.15	0.02	94.7
Cr	0.55	0.07	0.17	0.29	0.16	0.65	0.22	91.4
Cu	0.24	0.54	0.22	0.16	0.09	0.55	0.09	74.0
Ni	0.27	0.08	0.02	0.35	0.16	0.72	0.16	77.8
Br	0.16	0.05	0.15	0.19	0.08	0.11	0.84	80.8
I	0.15	0.24	0.12	0.06	0.05	0.13	0.81	77.7
Total variability (%)	30.3	11.7	10.2	9.8	8.3	8.0	5.7	84.0
Eingen value	17.4	3.62	2.88	2.46	1.76	1.34	1.10	

As dominant lithogenic marker for atmospheric distribution for the whole territory of the Macedonia was identified the following association: Al-As-Ba-Ce-Co-Cs-Dy-Fe-Hf-La-Li-Mg-Na-Nd-Sc-Sm-Ta-Tb-Th-Ti-U-V-W-Yb-Zr (Barandovski et al., 2015). Content of these elements in mosses is significantly influenced by secondary sources-mineral particles released into the atmosphere by wind erosion of local sources or particles attached to the moss in the periods when the soil surface is covered by water. The highest contents of the aforementioned group of elements have been found in moss samples that were collected on the Precambrian and Paleozoic schists or close to them (Figure 7). As a result of the actions of natural conditions, weathering of shales, very fast and consequently releases these elements into the environment occur.

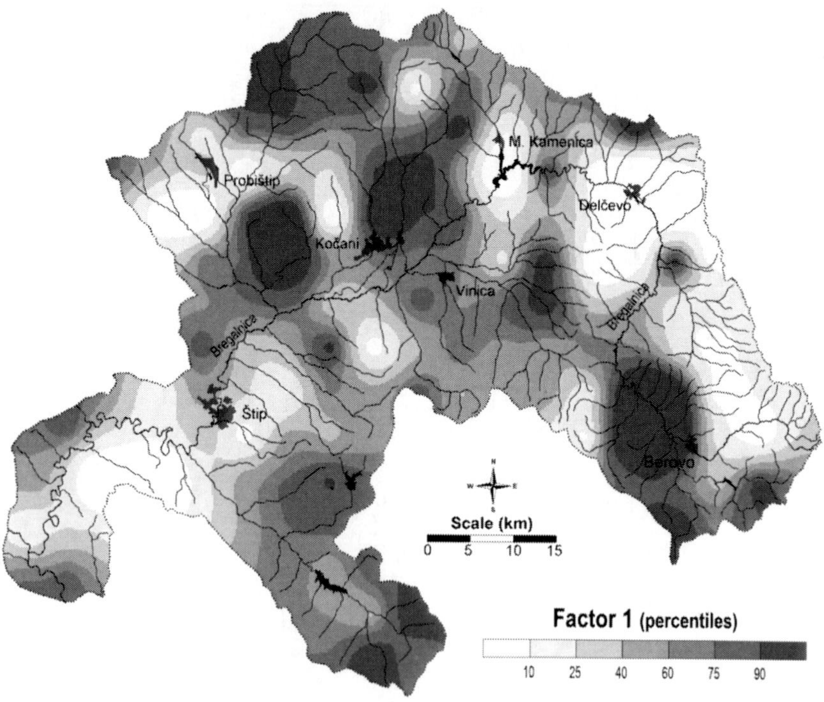

Figure 7. Areal distribution of Factor 1: Al-Be-Co-Fe-Ga-Ge-Li-Mg-Sc-Ti-V-Y-(La-Gd)-(Eu-Lu) in moss.

Factor 2 (Ba-Bi-Cd-Pb-Sb-Zn) is probably associated with long-range transport of air pollution and contributes to 11.7% of the variance. The highest loading values were obtained for Cd (0.88), Zn (0.82) and Pb (0.81). The elements Cd, Pb, Zn and Sb were also identified as anthropogenic marker for the whole territory of the R. Macedonia (Barandovski et al., 2015). The atmospheric deposition of these elements has been intensified in the areas of Neogene pyroclastite and Proterozoic and Paleozoic schists (Figure 8). There are two dominant sub-areas that mostly affect on the anthropogenic enrichment of these elements. The first one is an area with dominant occurrence of the Neogene pyroclastites and vulcanite's, where the Pb-Zn hydrothermal exploitation is operated ("Zletovo" mine). The second one is in the area of very old volcanic minerals (Paleogene volcanic sedimentary rocks, and Paleogene flysh) where the Pb-Zn hydrothermal exploitation is operated also ("Sasa" mine). This area of very old volcanic rocks continuous in NW direction where also occur a Pb-Zn mineral exploitation ("Toranica" mine).

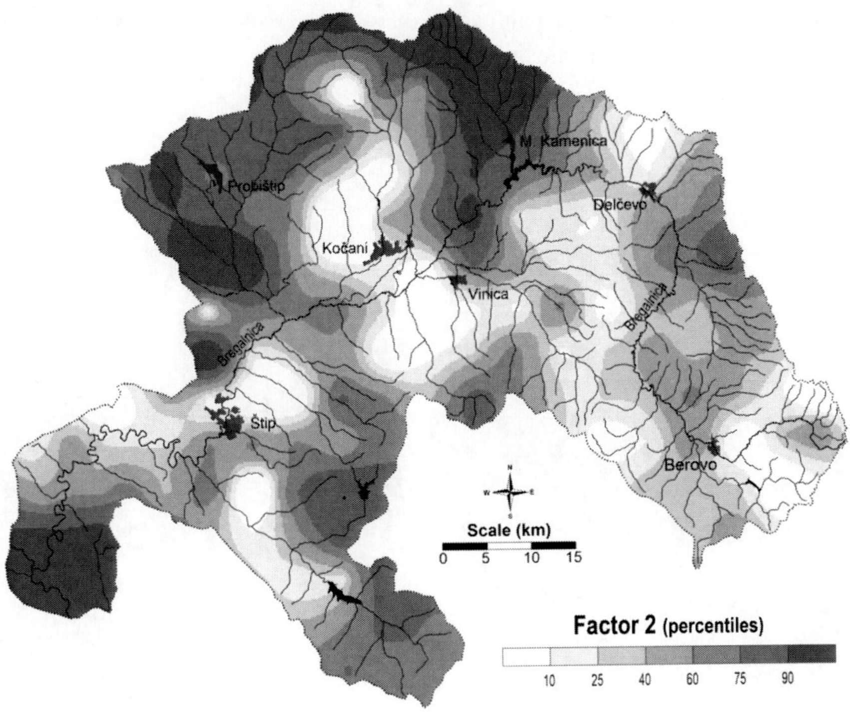

Figure 8. Areal distribution of factor 2: Ba-Bi-Cd-Pb-Sb-Zn in moss.

These mentioned sub-areas are predominantly affected with the dispersion of particles from flotation tailings dam sites near the "Zletovo" and "Sasa" mines. It should be pointed out that dust particles are significantly enriched with Pb and Zn contents. Soil surface wind-blowing dust is also affecting on the Pb-Zn enrichments, due to the significant soil contamination that occurs in these areas (Stafilov et al., 2014). Similar anomaly was recorded in northern France, exhibiting similar pattern for environmental distribution of Pb *vs.* Zn (Buorennane et al., 2010). This anomaly was extended with the Cd enrichments, due to the existence of the urbanized areas (Buorennane et al., 2010). Present data pointing to pollution source ("Zletovo" and "Sasa" mines). Lead and zinc distribution patterns were separately given, pointing on diagnosis of industrial pollution (Figure 8). The affected areas with significantly enriched Pb/Zn deposition predominantly were detected in the area of Neogene and Paleogene volcanism. The geochemistry of the igneous rock is known to correlate with extended poly-metallic enrichments (Keller et al., 2015). Furthermore, this association can be used as a very intensive probabilistic model of anthropogenic poly-metallic enrichments.

Factor 3 (As-Cs-Rb-Tl) indicates on characteristic deposition of these elements linked to old volcanism in the test area (Figure 9).

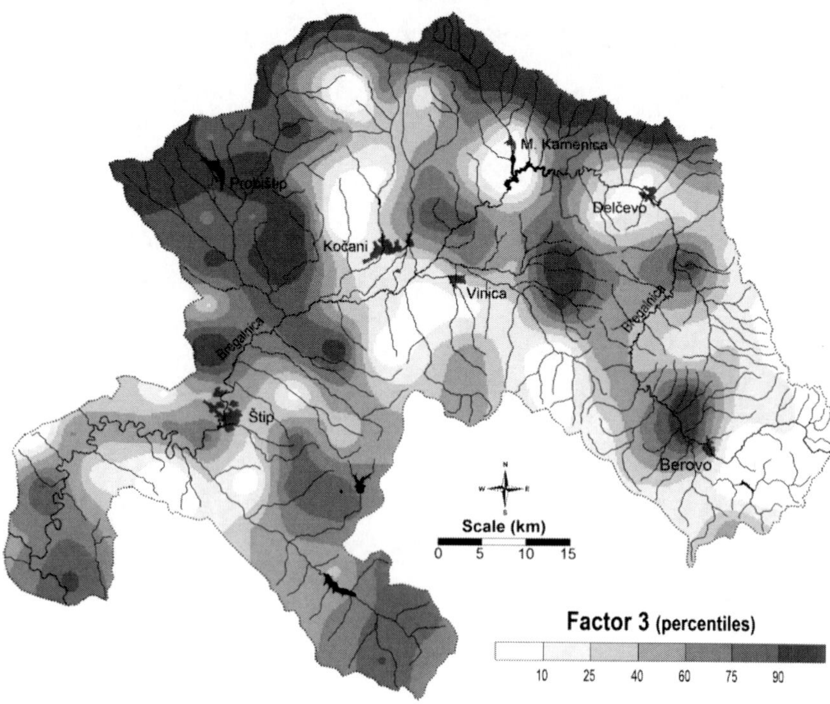

Figure 9. Areal distribution of factor 3: As-Cs-Rb-Tl in moss.

Spatial distribution of these elements follows the areal distribution of Factor 2. This geogenic marker probably occurs as secondary poly-metallic enrichments in area where the vulcanite dominates (Figure 9). Predominantly this occurrence marked the areas with Neogene vulcanite's vs. Paleogene vulcanite's.

The specific distribution of As, Cs, Rb and Tl was extended in the SE direction in the Berovo town vicinity. There are data that suggest a poly-metallic enrichment in this area (Serafimovski and Aleksandrov, 1995). This association can be used for the identification of the anthropogenic hydrothermal mineral exploitation, due to the areal distribution patterns (Figure 9). Higher paterns for these elements were obtained in Pb-Zn Sasa mine area, Pb-Zn Zletovo mine environ and copper mine Bučim area.

Factor 4 (Ca-Sr-B) was extracted with total variability of 9.8%. Distribution of Ca and Sr appear in the rocks of the Earth's crust, mainly as insoluble carbonates and sulfates and silicates (Greenwood and Earnshaw 2005). The processes of its total representation depend on the geochemical model they use, particularly the relative weights given to different types of volcanic and sedimentary rocks. According to the distribution in different geological units, this association dominates in Paleogene flysh (Figure 10).

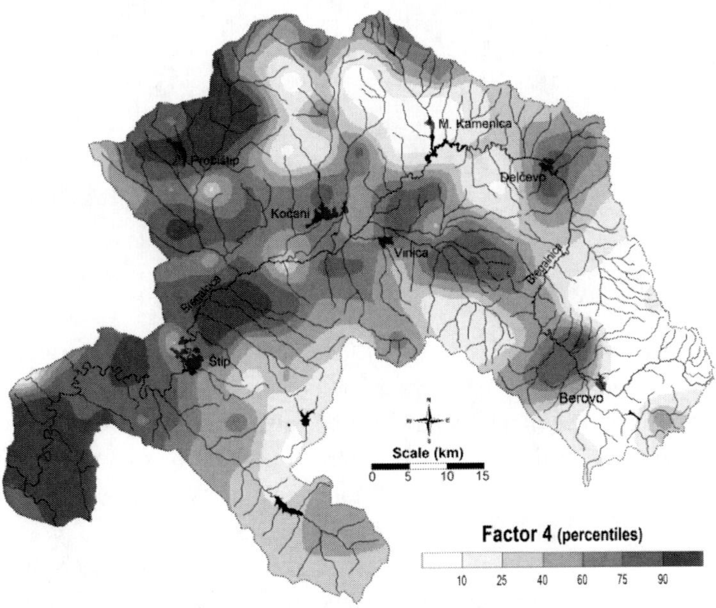

Figure 10. Areal distribution of factor 4: Ca-Sr-B in moss.

These old sedimentary rocks are natural archive for these elements. Along the whole couce of the Bregalnica river was determined dominance of this factor, with emphasis on the area where the Bregalnica river flows into Vardar river (intensive occurrence of Paleogene flysh, Figure 10). This lithogenic marker can be used for identification of natural enrichments due to the wind dusting in area of very old sedimentary rock in river basins.

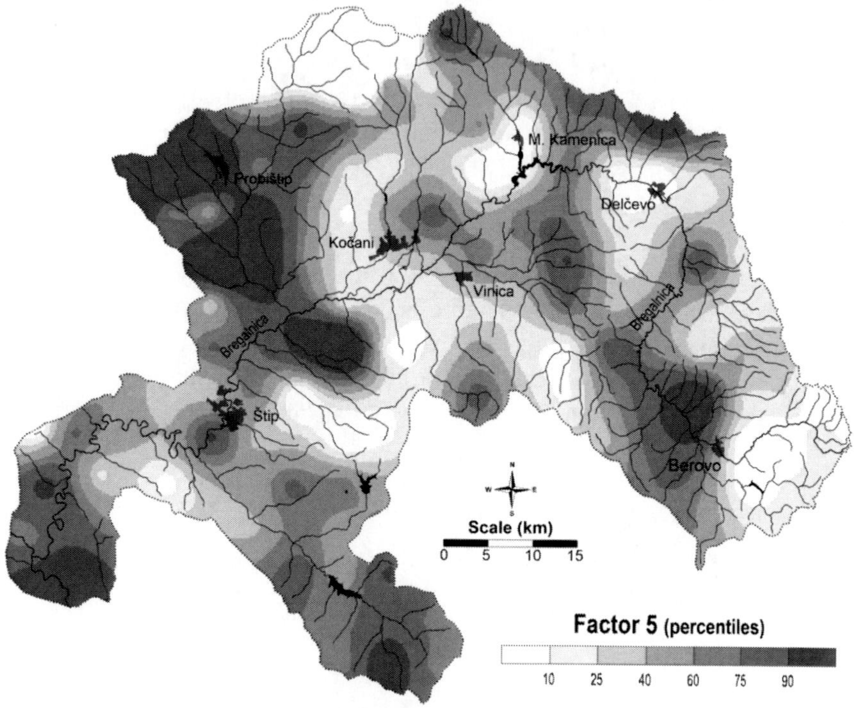

Figure 11. Areal distribution of factor 5: Hf-Zr in moss.

Factor 5 (Hf-Zr) presents a natural phenomenon in the investigated area, due to the specific geochemistry of these elements in environment (Figure 11). Their very similar chemical background correlated these elements. Zr and Hf behavior is similar to the HREEs (Prudêncio 2006). No significant variation occurs accordingly to dominant lithological units, but some intensive deposition can be singled out in area of Neogene volcanism. This area was characterized with poly-metalic enrichments (with special emphasis on Pb, Zn, Sb and Cd). Lower enrichments of Hf and Zr in this kind of area were also identified in Portugal, in correlation with significant Zn enrichments Prudêncio (2006). Barandovski et al., (2015) reports for decreasing trend in Hf-Zr distribution pattern from 2005 to 2010 on the whole territory of the Republic of Macedonia. Data obtained in this work compared to values from the investigation in 2010 (Barandovski et al., 2015) shows lower medians. Despite the decreasing trend in Hf-Zr deposition, still this geogenic association can be correlated to some poly-metallic anomalies (Figure 11).

Factor 6 (Cr-Cu-Ni) associate elements primarily affected by natural factors such as lithological background. Paleogene flysch sediments and Neogene sediments which occur along Vardar zone repeatedly demonstrated in Macedonia (Stafilov et al., 2010). These elements can be connected to air pollution as well. Most emissions of these elements to the environment originate from local sources, mainly from mines. Copper mine Bučim, contributes in the anthropogenic copper emission, in the South-East part of the study area. But, significant enrichment of this geochemical association occurs in the area of Paleogene flysch, intercepting a part from the Vardar river basin. Barandovski et al., (2015) identify a strong anomaly in Cr-Co-Ni distribution along the Vardar river basin. This anomaly affects a part from the Bregalnica river basin that was demonstrated also within this study (Figure 12).

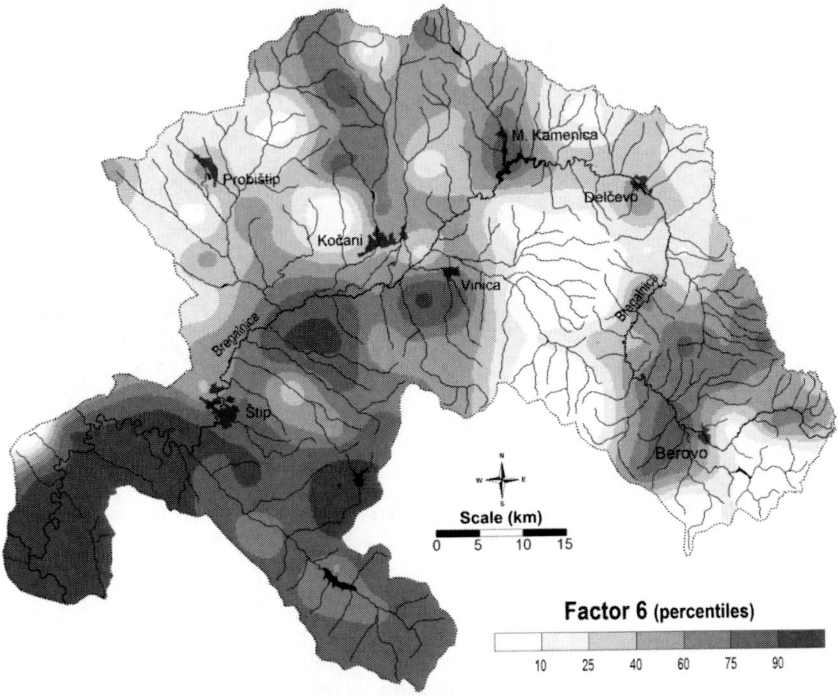

Figure 12. Areal distribution of factor 6: Cr-Cu-Ni in moss.

Factor 7 (Br-I) associates elements primarily affected by marine influence (Frontasyeva and Steinnes, 2004). Coast line of the Aegean Sea is less then 60 km from the south state border. Barandovski et al., (2015) reported for the influence of the wind blows along the Vardar river valley from the northern parts of Africa across the Mediterranean and the Aegean Sea. Elevated values of these elements in the mosses collected in area connected with the presence of Neogenic sediments along the river Begalnica (Figure 13). Lower enrichments of these elements were found in area of Rifeous schist.

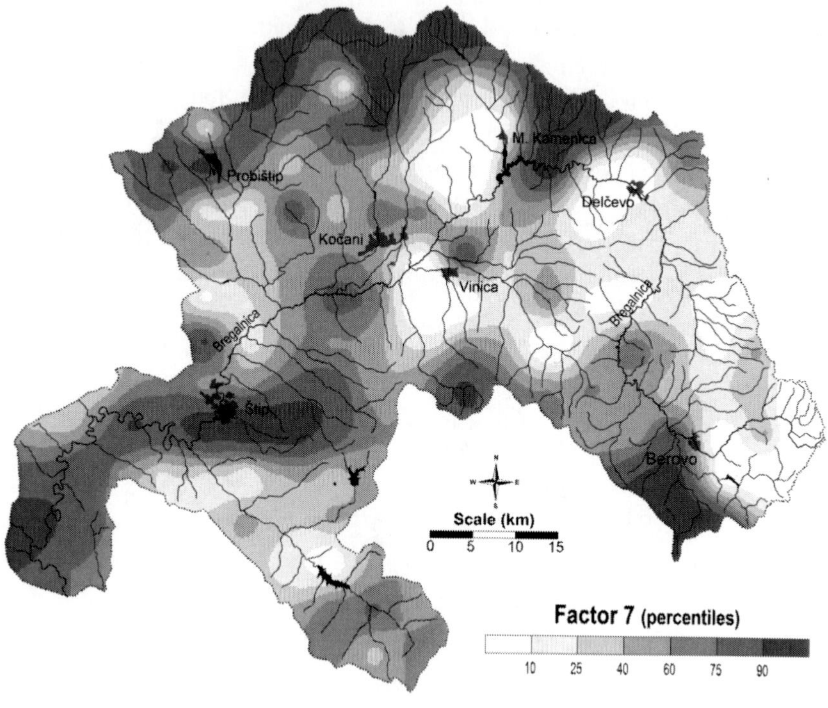

Figure 13. Areal distribution of factor 7: Br-I in moss.

4.2. Monitoring Distribution of Elements in Attic Dust Samples

4.2.1. Summary Results of the Data Set

From total 69 analyzed elements, for 60 elements were obtained values above the determined instrumental detection limits. Table 5 reports the descriptive statistic parameters (median and ranges values) corresponding to sixty elements contents in attic dust and topsoil samples from the house yards.

Data filtering was conducted to upgrade this investigation. The contents of the major elements in attic dust samples varied in thefollowing ranges: Al (4-7.4%), Ca (0.58-8.6%), Fe (0.07-16%), K (0.017-4.1%), Mg (0.08-2.6%), Na (0.01-2.6%), P (0.04-0.40%), Ti (0.002-0.79%). Almost identical results were obtained for the elements contents in topsoil samples, only for Ca and Fe were obtained extended distribution (max. value 17 % and 47%, respectively). Enrichments of these elements in Paleogene flysch areas suggest the typically depositions during an early stage of the orogenesis. The natural abundance of the major elements is strongly correlated to the sedimentary rocks in the investigated area. Keller et al., (2015) explain about similar anomaly composition rarity, during the volcanic-plutonic parity investigations. Enrichments in volcanic Ti and K may also suggest the influence of peralkaline volcanics onto the geochemistry of these elements. Predominantly, for the major elements distribution is the affect of the soil surface wind-blow dusting. The investigated area present a unique area of interweaves of the widely distributed type of rocks formed by high-grade regional metamorphic processes. More dynamic differences between lithogenic distributions in soil vs. elemental dust deposition occur for trace elements.

Trace elemental abundances consistently diverge in zones with occurrence of different lithological units. Rb, Cs, Sr, and Ba, are often collectively termed the large-ion-lithophile (LIL) elements. Because of their solubility, they are quite mobile during metamorphism and weathering. Thus the factors that most govern their behavior in igneous rocks are ionic radius and charge. Their enrichments in the investigated area were also correlated with intensive depositions in sub-areas with dominant occurrence of Neogene pyroclastite. In contrast to the heavy alkaline earths, Be and Li shows some different trends in chalcophile/lithophile deposition. River sediments areas dominates in Be and Li distribution where the maximum contents were deposited (3.4 and 44 mg/kg, respectively).

As the alkalis and alkaline earths, the rare earths and Y are strongly electropositive; thus have very similar geochemistry in the environment. Their abundance is usually correlated many continental rocks, as well as mostsediment. Many sedimentary rocks, andseawater, have rare earth patternsthat are similar to eachother, and to that of the continentalcrust. To accentuate the difference in rare earth patterns between sediments, low temperature geochemists often normalizer are earth abundances to the concentrations in average shale. Because the rare earths are highly insoluble and immobile, rare earth patterns often remain unchanged during metamorphism. Hence rare earth patterns can provide information on the premetamorphic history of a rock. Indeed, even during the production of sediment from crystalline rock, the rare earth patterns often remain little changed, and rare earth patterns have been used to identify the provenance, i.e., the source, of sedimentary rocks. Shale-normalized REE earth patterns of the Bregalnica river basin are as follow: La (18 mg/kg), Ce (39 mg/kg), Pr 4 mg/kg), Nd (14 mg/kg), Sm (2.8 mg/kg), Eu (0.68 mg/kg), Gd (3.6 mg/kg), Tb (0.46 mg/kg), Dy (2.6 mg/kg), Ho (0.5 mg/kg), Er (1.4 mg/kg), Tm (0.20 mg/kg), Yb (1.3 mg/kg), Lu (0.19 mg/kg). Characterization of REE distribution in authomorphic and alluvial soil, reveals that the sum for total light rare earth elements (ΣLREEs) in the Bregalnica river basin ranges of 8.6-225 mg/kg, while for the total sum of the heavy rare earth elements (ΣHREEs) ranges from 0.92-33.7 mg/kg (Balabanova et al., 2015).

Zr and Hf are nominally incompatible in all major rock-forming minerals. Keller et al., (2015) report about inflection decreasing trends of Zr and Hf with increasing silica reflects saturation of zircon ($ZrSiO_4$). Rubatto (2002) and Kabata-Pendias (2010) observed these elements tend to be very immobile during weathering andmetamorphism. They are therefore particularly valuable in the study of ancient igneous rock suites asthey can sometimes provide insights into the environment in which those rocks formed. These elements range from 0.08 to 3.0 mg/kg for Hf and from 4.9 to 96 mg/kg. They are therefore particularly valuable in the study of igneous rock suites asthey can sometimes provide insights into the environment in which those rocks formed. Enriched median values were obtained in the Paleogene and Neogene volcanism. Zirconium was enriched to 27 mg/kg for both sub-areas, hafnium was enriched to 0.68 mg/kg.Ta and Nb are present in anomalously low concentration sin magmas associated with subduction zones; indeed, this is considered a diagnostic feature of subduction-related volcanism (Keller et al., 2015). Although this depletion is not well understood, it is probably at least in part a consequence of the low solubility of these elements and the consequent failure of aqueous fluids generated by dehydration of the subducting oceanic crust totransport these elements into the magma genesis zone Keller et al., 2015). Nb content in air deposited dust ranges from 0.28-17 mg/kg, while in top soil layer riches to 31 mg/kg. Similar distribution was obtained for Ta, from 0.05-1.3 mg/kg in attic dust and from 0.14-2.5 mg/kg in top soil (Table 5).

The geochemistry of the transition elements is considerably more complex than that of the elements we have discussed thus far. Their behavior in magmas is also variable. They range from moderately incompatible (e.g., Cu, Zn) to very compatible (e.g., Cr, Ni), but their exact behavior is generally a stronger function of composition than that of the highly incompatible elements. Stafilov et al., (2012) reports about anthropogenic anomaly of Cu deposition Cd, Cr, Cu, Ni, Pb and Zn in mine environ (max values: 3.1, 110, 415, 59, 121 and 93 mg/kg, respectively). Balabanova et al., (2012) also explains that urban elements and those related to mine activities in dust were associated with fine particles, which are preferentially mobilized by wind and trapped in attics. Present investigation also demonstrates a very similar anomaly for abovementioned elements. Copper content in air-distributed dust ranges from 6.7 to 880 mg/kg; while in topsoil layer an extended distribution has occurred (max. value 1200 mg/kg). Enriched values were obtained from houses located very close to Cu mine Bučim. Cadmium contents also hsave an enrichments trend in air-distributed dust (0.054-25 mg/kg). Despite the Bučim area, enriched Cd (>10 mg/kg) contents were obtained in Pb-Zn Sasa mine. Content of Cd in topsoil layer does not indicate any significant enrichment (0.005-9 mg/kg). Therefore, the attic dust depositions indicate on the long time anthropogenic affected distribution in mine environs. Agarwal (2009) explain Cd enrichments of topsoil horizons is ascribed to agricultural activities such as the use of urban sludge, manure and phosphate-enriched fertilizers that are known to contain Cd in addition to atmospheric deposition in any case. Very similar to Cu and Cd, distribute Pb and Zn in the investigated area. Lead content ranges from 0.005 to 3900 mg/kg, while Zn contents shows lower enrichments from 26 to 3200 mg/kg. Lithogenic distribution in soil riches to 1200 mg/kg and 590 mg/kg, respectively for Pb and Zn. These significantly enriched contents for Pb and Zn are correlated with the lithogenic dominance of Rifeous shales, where the median values for Pb and Zn contents are 110 an 220 mg/kg, respectively. As most affected area with Pb/Zn enrichment separate Sasa mine environ, due to the continually long time wind dust dispersion from the flotation tailings.

From the platinum group, Pt and Pd were determined in traces above the instrumental limit detection. Their content distributes almost identically in attic dust and in surface soil. Platinum ranges from 0.005 to 0.78 mg/kg in soil, while in dust riches to 0.81 mg/kg. Palladium is more dominant platinum element, and distributes from 0.05 mg/kg to 1.4 mg/kg in attic dust and 1.7 mg/kg in soil. White (2013) clarify that chromites and sulfides are highly enriched in noble metals relative to the silicate magmatic rocks, consistent with their chalcophile nature and demonstrated affinity for oxide phases such as the spinels.

Table 5. Data summary for elements contents in attic dust samples (D) and topsoil samples (T); N = 168

Element	Unit	Median	Min	Max	Element	Unit	Median	Min	Max
Ag (D)	mg/kg	0.54	0.005	3.0	Mg (D)	%	0.59	0.08	2.6
Ag (T)	mg/kg	0.45	0.22	4.6	Mg (T)	%	0.47	0.06	7.2
Al (D)	%	5.7	4.0	7.4	Mn (D)	mg/kg	530	115	6900
Al (T)	%	6.0	3.0	8.9	Mn (T)	mg/kg	560	179	7000
As (D)	mg/kg	17	4.1	150	Mo (D)	mg/kg	0.67	0.005	17
As (T)	mg/kg	12	4.0	93	Mo (T)	mg/kg	0.48	0.005	8.3
B (D)	mg/kg	11	0.005	63	Na (D)	%	0.48	0.01	2.6
B (T)	mg/kg	0.56	0.005	42	Na (T)	%	0.78	0.08	3.3
Ba (D)	mg/kg	400	6.8	1400	Nb (D)	mg/kg	5.8	0.28	17
Ba (T)	mg/kg	420	76	2600	Nb (T)	mg/kg	5.8	2.4	31
Be (D)	mg/kg	0.93	0.13	3.4	Nd (D)	mg/kg	13	0.37	42
Be (T)	mg/kg	1.1	0.12	4.6	Nd (T)	mg/kg	12	3.1	69
Bi (D)	mg/kg	0.30	0.005	5.9	Ni (D)	mg/kg	21	4.4	250
Bi (T)	mg/kg	0.19	0.046	1.6	Ni (T)	mg/kg	14	4.6	260
Br (D)	mg/kg	2.4	0.005	20	P (D)	%	0.21	0.04	0.40
Br (T)	mg/kg	1.6	0.005	15	P (T)	%	0.07	0.01	0.31
Ca (D)	%	3.9	0.58	8.6	Pb (D)	mg/kg	51	0.005	3900
Ca (T)	%	2.3	0.26	17	Pb (T)	mg/kg	22	0.005	1200
Cd (D)	mg/kg	0.43	0.054	25	Pd (D)	mg/kg	0.35	0.05	1.4
Cd (T)	mg/kg	0.13	0.005	9.0	Pd (T)	mg/kg	0.31	0.05	1.7
Ce (D)	mg/kg	39	0.80	130	Pr (D)	mg/kg	3.6	0.08	11
Ce (T)	mg/kg	36	8.2	160	Pr (T)	mg/kg	3.5	0.82	18
Co (D)	mg/kg	6.6	0.94	20	Pt (D)	mg/kg	0.21	0.005	0.81
Co (T)	mg/kg	6.7	2.5	47	Pt (T)	mg/kg	0.16	0.005	0.78
Cr (D)	mg/kg	48	11	270	Rb (D)	mg/kg	48	3.5	120
Cr (T)	mg/kg	38	3.8	330	Rb (T)	mg/kg	54	9.2	250
Cs (D)	mg/kg	2.3	0.005	12	Sb (D)	mg/kg	0.63	0.060	3.2
Cs (T)	mg/kg	2.6	0.35	27	Sb (T)	mg/kg	0.42	0.10	3.9
Cu (D)	mg/kg	30	6.7	880	Sc (D)	mg/kg	6.9	0.54	92
Cu (T)	mg/kg	17	4.1	1200	Sc (T)	mg/kg	6.2	1.1	26
Dy (D)	mg/kg	2.5	0.08	7.8	Sm (D)	mg/kg	2.8	0.05	8.0
Dy (T)	mg/kg	2.3	0.66	15	Sm (T)	mg/kg	2.5	0.65	14
Er (D)	mg/kg	1.3	0.03	4.0	Sn (D)	mg/kg	3.2	0.005	23
Er (T)	mg/kg	1.2	0.35	7.6	Sn (T)	mg/kg	1.5	0.005	14
Eu (D)	mg/kg	0.72	0.02	2.3	Sr (D)	mg/kg	140	6.0	720
Eu (T)	mg/kg	0.68	0.16	4.7	Sr (T)	mg/kg	110	23	760
Fe (D)	%	1.5	0.07	16	Ta (D)	mg/kg	0.46	0.05	1.3
Fe (T)	%	1.5	0.18	47	Ta (T)	mg/kg	0.54	0.14	2.5
Ga (D)	mg/kg	9.1	2.1	20	Tb (D)	mg/kg	0.47	0.02	1.3
Ga (T)	mg/kg	9.2	5.3	56	Tb (T)	mg/kg	0.43	0.12	2.7
Gd (D)	mg/kg	3.6	0.13	10	Te (D)	μg/kg	31	5.0	1300

Element	Unit	Median	Min	Max	Element	Unit	Median	Min	Max
Gd (T)	mg/kg	3.3	0.95	21	Te (T)	µg/kg	17	5.0	280
Ge (D)	mg/kg	0.57	0.27	2.1	Ti (D)	%	0.23	0.002	0.79
Ge (T)	mg/kg	0.62	0.40	2.6	Ti (T)	%	0.22	0.03	0.96
Hf (D)	mg/kg	0.63	0.02	1.9	Tl (D)	mg/kg	0.34	0.05	1.7
Hf (T)	mg/kg	0.56	0.08	3.0	Tl (T)	mg/kg	0.38	0.05	2.5
Ho (D)	mg/kg	0.46	0.01	1.5	Tm (D)	mg/kg	0.18	0.01	0.52
Ho (T)	mg/kg	0.43	0.12	2.8	Tm (T)	mg/kg	0.17	0.05	1.1
I (D)	mg/kg	0.14	0.005	0.53	V (D)	mg/kg	72	29	370
I (T)	mg/kg	0.11	0.02	0.71	V (T)	mg/kg	62	28	600
In (D)	µg/kg	42	5.0	710	W (D)	mg/kg	1.0	0.045	8.9
In (T)	µg/kg	31	11	190	W (T)	mg/kg	0.99	0.24	5.9
K (D)	%	0.83	0.017	4.1	Y (D)	mg/kg	11	0.44	30
K (T)	%	0.89	0.16	6.1	Y (T)	mg/kg	8.9	2.4	58
La (D)	mg/kg	18	0.37	52	Yb (D)	mg/kg	1.1	0.050	3.4
La (T)	mg/kg	16	3.6	72	Yb (T)	mg/kg	1.1	0.33	7.1
Li (D)	mg/kg	13	3.4	44	Zn (D)	mg/kg	99	26	3200
Li (T)	mg/kg	13	2.8	65	Zn (T)	mg/kg	53	18	590
Lu (D)	mg/kg	0.17	0.005	0.56	Zr (D)	mg/kg	24	0.42	61
Lu (T)	mg/kg	0.17	0.05	1.1	Zr (T)	mg/kg	21	4.9	96

4.2.2. Multivariate Assessing for Dominant Geochemical Markers

Multivariate extraction was applied for reducing the numerous elements distribution and identifying the dominant geochemical association. On the basis of the matrix of correlation coefficients, multivariate cluster and R-mode factor analysis was used to reveal associations of chemical elements and to decrease the number of variables for the obtained data. Extraction was simplified to 6 factor association, marked as: F1: Ga-Nb-Ta-Y-(La-Gd)-(Eu-Lu), F2: Be-Cr-Li-Mg-Ni; F3: Ag-Bi-Cd-Cu-In-Mn-Pb-Sb-Te-W-Zn, F4: Ba-Cs-Hf-Pd-Rb-Sr-Tl-Zr; F5: As-Co-Ge-V, F6: K-Na-Sc-Ti, with a total variability for dominant loadings of 81.5%. Table 6 summaries factor loadings for dominant elements associations. Variables with factor loading higher than 0.5 were assumed to contribute significantly to a given factor. The remaining elements were eliminated from the analysis, due to their tendency to form separate clusters, thus not showing a reasonable connection with other chemical elements and to satisfy the criteria of dimension variables based on number of observations. Elements with a low share of communality or tendency to form independent factors were also excluded. Factor analysis (FA) was performed on elements standardized to zero mean and unit standard deviation. The inter-elements similarity and distancing were revealing using the cluster dendrogram (Figure 14). Most distant clusters were Factor 1 [Ga-Nb-Ta-Y-(La-Gd)-(Eu-Lu)] and Factor 3 (Ag-Bi-Cd-Cu-In-Mn-Pb-Sb-Te-W-Zn).

Table 6. Matrix of factor loadings-factor analysis (FA) of the elements' contents

Element	F1	F2	F3	F4	F5	F6	Communality
Ga	0.64	0.15	0.06	0.29	0.59	0.16	89.9
Nb	0.80	0.17	0.18	0.22	-0.02	0.27	82.8
Ta	0.77	0.07	0.16	0.42	-0.08	0.19	84.5
Y	0.87	0.17	0.14	0.11	0.00	0.21	86.4
La-Gd	0.81	-0.02	0.12	0.51	0.00	0.04	92.8
Eu-Lu	0.79	0.03	0.11	0.40	0.04	0.19	83.6
Be	0.11	0.53	0.41	0.10	0.25	0.42	71.1
Cr	0.24	0.83	-0.06	-0.10	0.23	0.20	85.4
Li	-0.06	0.76	0.22	0.33	0.10	0.18	77.8
Mg	0.10	0.84	0.17	-0.13	0.06	0.37	90.5
Ni	0.09	0.89	0.09	0.05	0.11	0.00	82.7
Ag	0.31	0.05	0.73	0.13	0.23	0.10	71.1
Bi	0.23	0.01	0.88	0.09	-0.09	0.03	84.7
Cd	-0.06	-0.01	0.82	0.36	0.08	0.08	81.6
Cu	0.26	0.32	0.64	-0.06	0.08	0.20	63.7
In	0.40	-0.18	0.70	0.31	0.21	0.08	82.8
Mn	0.27	0.20	0.68	0.31	0.35	0.21	83.3
Pb	-0.11	0.35	0.77	-0.05	-0.13	0.24	80.6
Sb	-0.13	0.16	0.77	0.38	0.00	0.14	79.9
Te	0.02	0.23	0.73	0.04	0.31	-0.13	70.2
W	0.38	0.08	0.85	0.20	-0.01	0.01	90.7
Zn	-0.11	-0.03	0.74	0.18	0.30	0.24	74.36
Ba	0.38	-0.26	0.05	0.79	0.00	0.10	84.2
Cs	0.20	0.21	0.31	0.76	-0.22	-0.17	83.9
Hf	0.34	0.00	0.12	0.86	0.01	0.14	87.9
Pd	0.05	-0.04	0.03	0.73	0.43	0.22	78.4
Rb	0.51	0.25	0.29	0.55	0.02	0.04	71.3
Sr	0.17	0.10	0.31	0.82	0.11	0.16	84.1
Tl	0.22	-0.03	0.51	0.72	-0.01	-0.08	83.9
Zr	0.30	0.12	0.15	0.84	0.07	0.16	86.9
As	-0.42	0.11	0.33	0.02	0.72	-0.15	83.5
Co	0.35	0.44	0.28	0.23	0.45	0.31	74.1
Ge	0.18	0.29	0.19	0.04	0.86	0.06	89.8
V	-0.23	0.30	0.10	-0.13	0.59	0.53	80.2
K	0.08	0.31	0.30	0.38	0.11	0.67	80.2
Na	0.28	0.18	0.11	0.10	0.15	0.84	85.9
Sc	0.29	0.14	0.11	0.19	-0.06	0.69	63.8
Ti	0.50	0.21	0.10	0.03	-0.03	0.76	88.7
Total variability (%)	15.6	11.3	20.1	16.8	8.03	9.61	81.5
Eingene value	15.2	5.36	4.36	2.46	1.91	1.71	

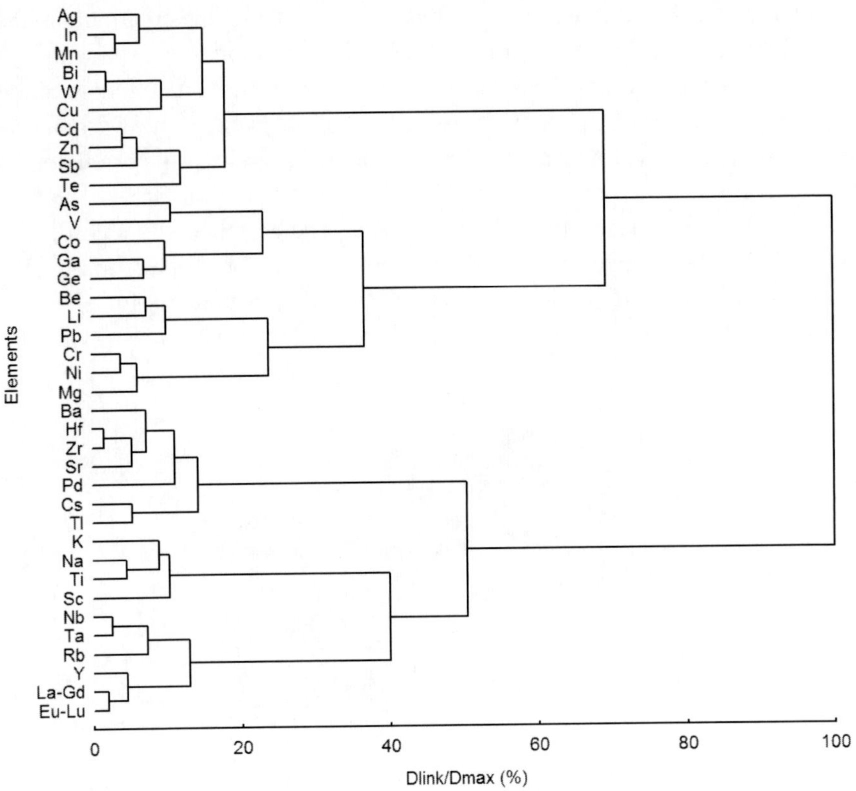

Figure 14. Dendrogram for dominant clusters.

4.2.3. Areal Distribution Mapping of the Factors

For the spatial distribution of the factor scores, universal kriging with the linear variogram interpolation method was applied for the construction of maps showing the spatial distribution of factor scores (F1-F6).

Distribution of Factor 1 [Ga-Nb-Ta-Y-(La-Gd)-(Eu-Lu)] clearly expresses the typical connection of some basic Earth forming process, due to the incorporation of the rare earth. Their sources are mainly natural phenomena such as rock weathering and chemical processes in soil. White (2013) explains that many sedimentary rocks have rare earth patterns that are similar to each other, and to that of the continental crust. However, the geochemistry of the rare earths in the investigate area doesn't dominate on the sedimentary rocks. Occurrence of this factor is typical for the Proterozoic formations (gneisses, granites and shale) (Figure 15). Because the rare earths are highly insoluble and immobile, rare earth patterns often remain unchanged during metamorphism. Hence rare earth patterns can provide information on the premetamorphic history of a rock. Indeed, even during the production of sediment from crystalline rock, the rare earth patterns often remain little changed, and rare earth patterns have been used to identify the provenance, i.e., the source, of sedimentary rocks. A unique occurrence is correlation of these elements with Ta and Nb. Tantalum and niobium are present in anomalously low concentrations in magmas associated with subduction zones (indeed, this is considered a diagnostic feature of subduction-related volcanism). As the Figure 15 visualizes the maximum patterns for elements deposition, Proterosoic gneisses were most dominant areas with F1 enrichments.

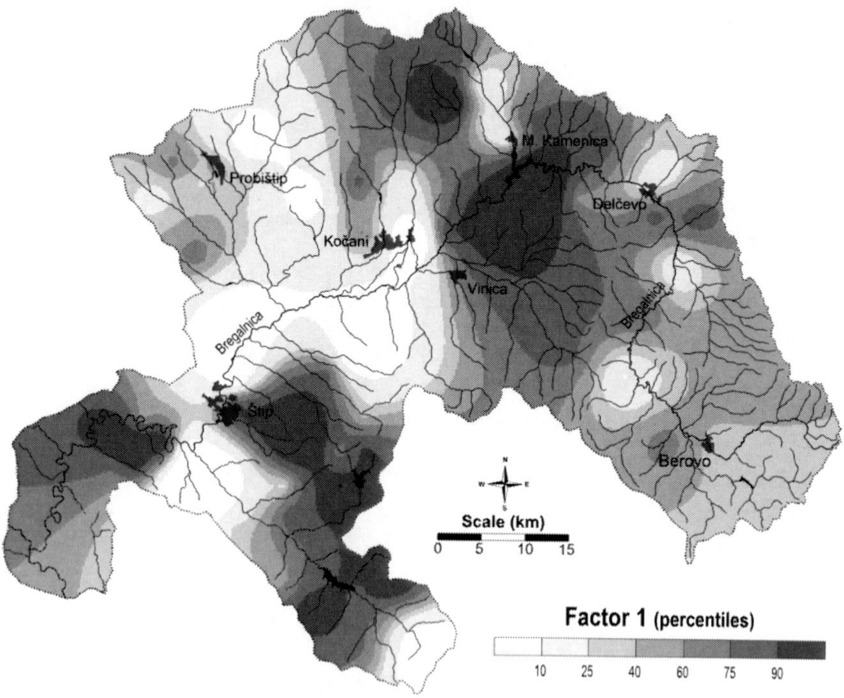

Figure 15. Distribution patterns for Factr 1: Ga-Nb-Ta-Y-(La-Gd)-(Eu-Lu).

The second geochemical association (Be-Cr-Li-Mg-Ni) relays also on the effect from the wind-blowing dust from the surface soil layers. These determined elements are considered "natural" because their origin is primarily crustal-soil particles suspended and transported by wind. High factor loadings are related to some old formation such as Paleogene flysch and Quaternary unconsolidated sediments (Figure 16). The anomaly of the Cr-Ni distribution was severally times improved along the Vardar zone repeatedly demonstrated in Macedonia (Stafilov et al., 2010; Balabanova et al., 2010; 2011; et al., 2013, 2015), and in Balkan countries. Higher contents of aforementioned elements found in distributed dust along the valleys of rivers Vardar and Crna, where constant flow of air masses in both direction is present (Lazarevski, 1993), indicate an influence of lithological background (Bačeva et al., 2012). The specific distribution of Be-Cr-Li-Mg-Ni was extended in the SE direction in the Berovo town vicinity. There are data that suggest a poly-metallic enrichment in this area (Arsovski, 1997). These elements can be connected to air pollution as well. Most emissions of these elements to the environment originate from local sources, mainly from mines and smelters (Stafilov et al., 2014). Furthermore, this association can be used for the identification of the poly-metallic anomalies as secondary occurrence.

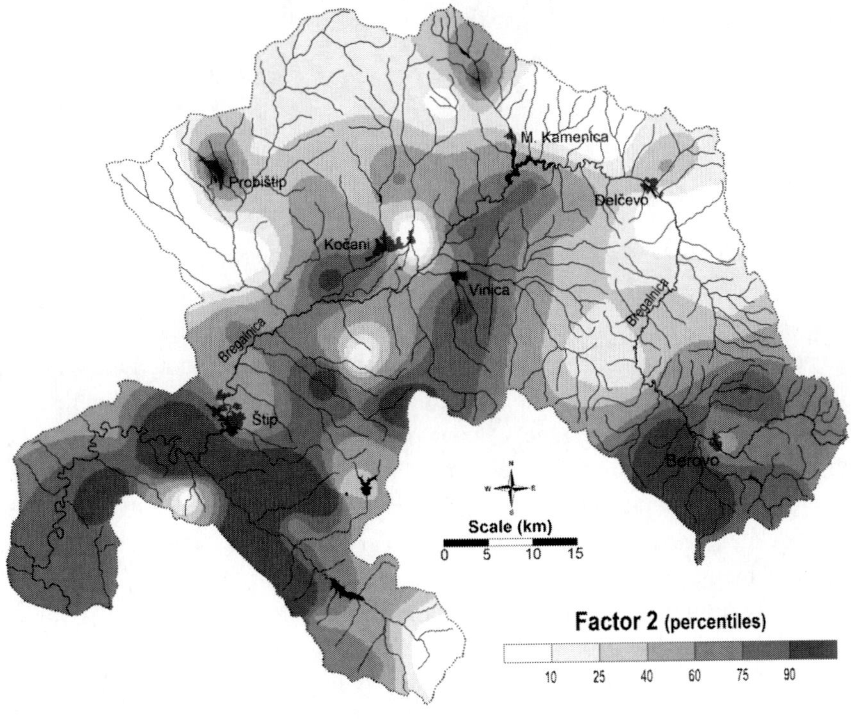

Figure 16. Distribution patterns for Factor 2: Be-Cr-Li-Mg-Ni.

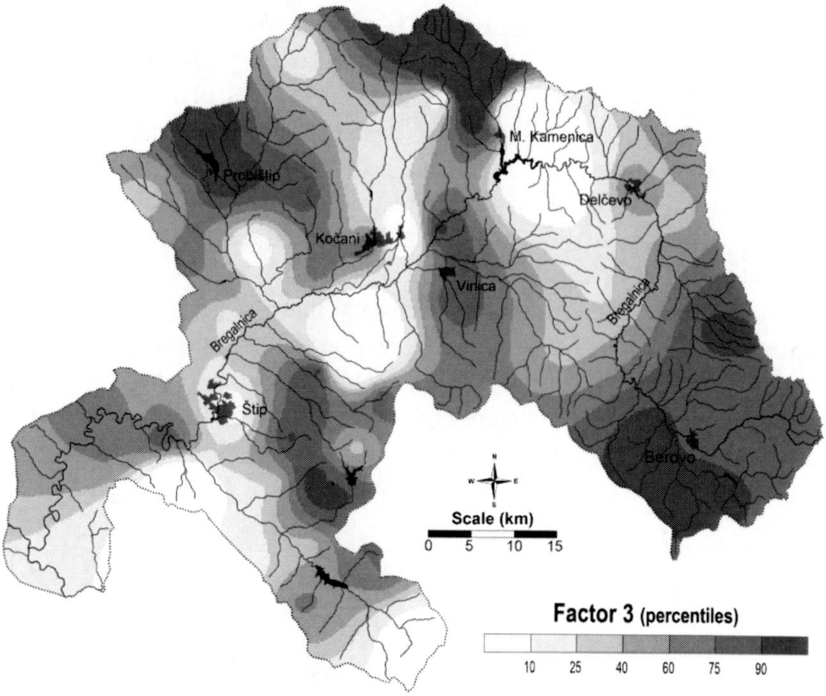

Figure 17. Distribution patterns for Factor 3: Ag-Bi-Cd-Cu-In-Mn-Pb-Sb-Te-W-Zn.

Factor 3 (Ag-Bi-Cd-Cu-In-Mn-Pb-Sb-Te-W-Zn) was the most heterogenic geochemical association. Enriched deposition occurred dominantly on Rifeous shale (Figure 17). These very old rocks occur as transition lithologic units between the Neogene and Paleogene volcanism in the investigated area. Pb-Zn mines "Sasa" and "Zletovo" are located in the areas with dominantly presence of the Oligocene and Neogene volcanism appears sequentially and in several phases forming sub-volcanic areas. On the other hand, the pyroclastites are most frequently found in the Kratovo-Zletovo volcanic area, where the dacites and andesites are the oldest formations. This geochemical association links typical elements which are normally associated with air pollution (Cd-Pb-Zn), and usually are not influenced by lithological background (Figure 17). Spatial patterns show intensive deposition in the area of poly-metallic hydrothermal exploitations, Sasa, Zletovo and Bučim (Figure 17). Thus, this factor can explain and mark the anthropogenic affects on air pollution. These anthropogenic anomalies have an historical record for decades continuously introduction of significant emission of dust particle with enriched contents of Cd, Pb and Zn. Also, this investigation fortifies an extended anthropogenic association (Ag, Bi, In and Mn) that implement and some other anthropogenic activities such as agricultural activities (use of urban sludge, manure and phosphate fertilizers) or their occurrence can be a secondary affection from mine poly-metallic pollution. Affiliation of Te and W to this group, spread up a novel poly-metallic/metalloid anomaly in the Berovo region. For all elements that associate in this group, enriched values deposition occur in this mention area. Almost twenty years ago, Arsovski (1997) point on poly-metallic enrichment in this area so called Vladimirovo-Berovo, during the tectonic investigation. The present investigation also interpolates this area as metallic's/metalloids enriched zone, with emphasis on the anthropogenic elements.

The occurrence of Factor 4 (Ba-Cs-Sr-Rb-Pd-Tl-Zr-Hf), presents an interesting geochemical association, incorporating the incompatible/ hygromagmatophile elements. Elements enrichments occur in areas with predominance of Neogene pyroclastite and sediments (Figure 18).

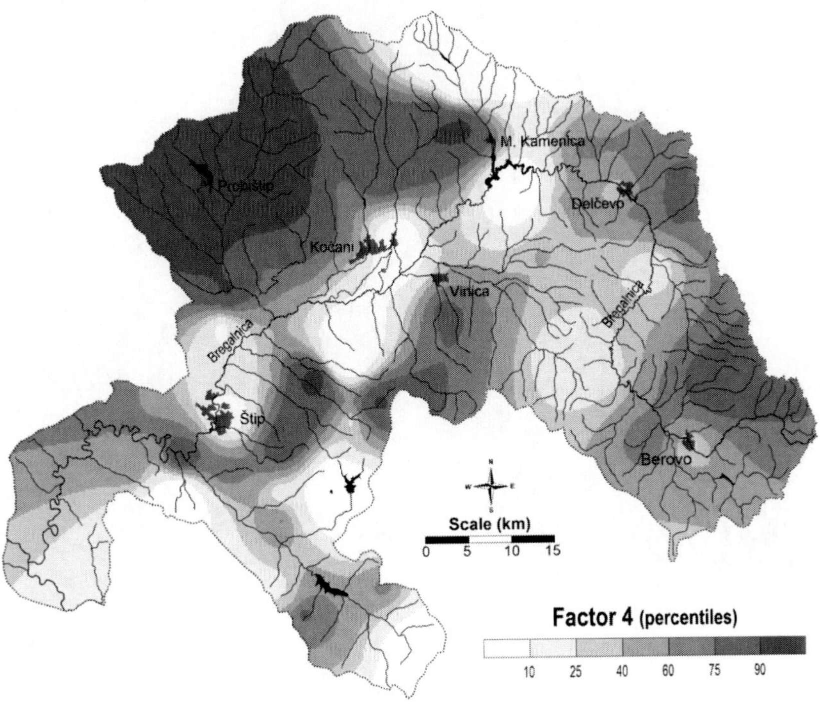

Figure 18. Distribution patterns for Factor 4: Ba-Cs-Hf-Pd-Rb-Sr-Tl-Zr.

According to generalized geology map (Figure 2), Kratovo-Zletovo region is the unique district in the region located along the continental margin and is closely related to the Tertiary volcanoes and hydrothermal activities in this area. According to Dumurdzanov et al., (2004) the pyroclastites are most frequently found in the Kratovo-Zletovo volcanic area, where the dacites and andesites are the oldest formations. These polyphasal Neogene deformations through the insignificant movements associated with the volcanic activities had direct influence on the gradual formation of the reefs and the formation of deposits in the Zletovo area. Spatial patterns are extended in eastern direction, due to the most common winds from western direction with frequency of 199‰ and speed of 2.7 m/s (Lazarevski, 1993). Barandovski et al., (2015) singled out lithogenic markers of Ba-Cs association enriched in areas of felsic volcanic rocks (andesite) and their pyroclastic rocks primarily occupies the Kriva Reka basin, bordered with the Kratovo-Zletovo region. The same geochemical anomaly occurs in smaller zones (Kožuf and Mariovo regions).

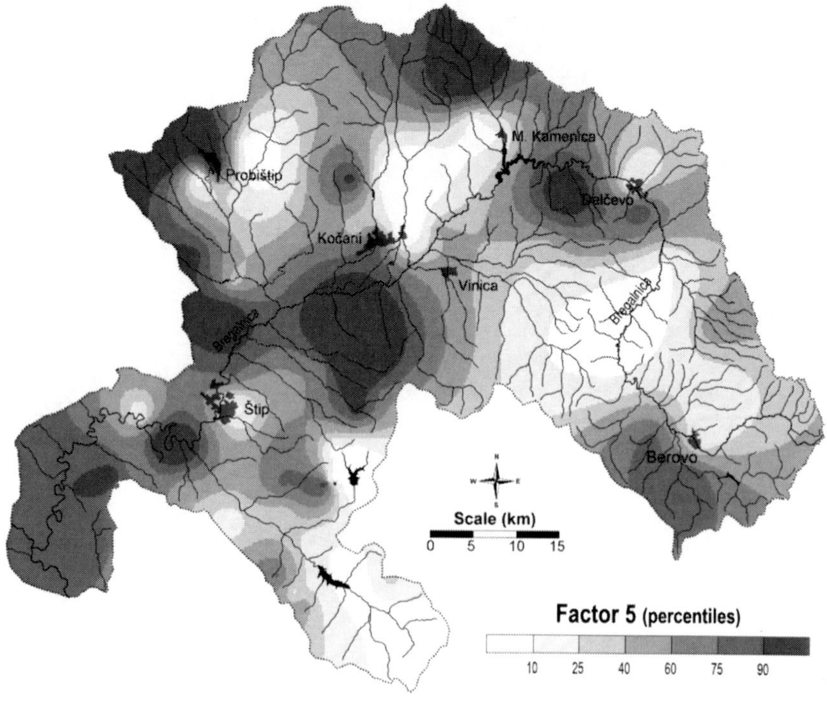

Figure 19. Distribution patterns for Factor 5: As-Co-Ge-V.

Factor 5 associates the following elements, As, Co, Ge and V, with predominant occurrence on Rifeous shales (Figure 19). This kind of geochemical *fingerprinting* occurs along the whole course of the Bregalnica river. Accordingly, the resulting areal distribution map used to support with high certainty the assessment for poly-metallic enrichments as ascribed to urbanization, including vehicular emissions and incinerators and industry. Comparative analysis for the areal patterns singled out a very similar behavior of the Factor 3 and Factor 5. Clustering method was very useful for determination of both geochemical associations. Dendrogram displays a very close interpolation of the mention elements from F3 and F5. Furthermore, there is a strong interconnection between the anthropogenic and lithogenic fingerprinting. Basically, the element geochemistry intermediate between atmospheric emissions and lithogenic wind-blow dusting. Therefore, the As-Co-Ge-V distribution can be used as proposed mechanism for possible tracking of anthropogenic poly-metallic enrichments.

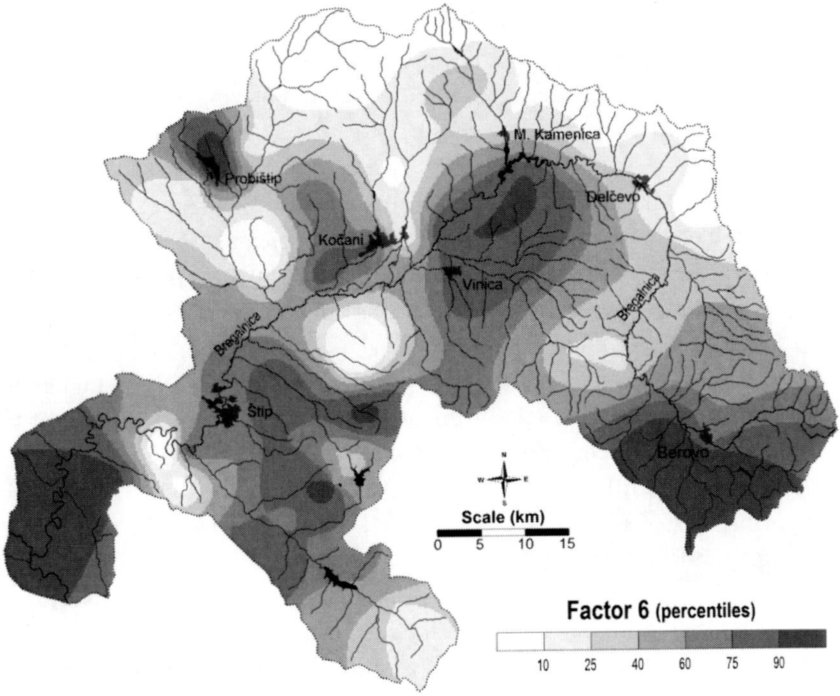

Figure 20. Distribution patterns for Factor 6: K-Na-Sc-Ti.

The sixth factor includes K-Na-Sc-Ti. Although it is the least distinguished factor; nevertheless this geochemical association is related with the lithogenic dominance of the Neogene sediments and Paleogene flysch as given in Figure 20. Their occurrence in the environment is usually correlated to the major minerals in basaltic and ultramafic rocks have two kinds of cationic lattice sites: small tetrahedral sites occupied by Si and Al (White et al., 2013). Because of their solubility, they are quite mobile during metamorphism and weathering. Basic hotspots were correlated with the sedimentary mineralization along the whole course of the Bregalnica river terraces. Quaternary terraces in Kočani valley are also marked with this lithogenic marker.

CONCLUSION

The focus of this investigation has been given on determinating and mapping of lithogenic and anthropogenic atmospheric distribution of 69 elements in the Bregalnica river basin, Republic of Macedonia, using moss and attic dust as sampling media. Using moss biomonitoring technique, several highlights could be point out:

- The specific poly-metallic enrichments occur in the areas of the very old Neogene volcanism and Paleogene volcanism. Strong relations occur between these dominant lithological units and the hydrothermal exploitation in the area of Pb-Zn Sasa mine, Pb-Zn Zletovo mine and Bučim copper mine.
- The multivariate approach and spatial distribution *hunting*, singled out seven dominant cluster for the atmospheric distribution of total of 69 analyzed elements: F1: Al-Be-Co-Fe-Ga-Ge-Li-Mg-Sc-Ti-V-Y- (La-Gd)-(Eu-Lu); F2: Ba-Bi-Cd-Pb-Sb-Zn; F3: As-Cs-Rb-Tl; F 4: Ca-Sr-B; F5: Hf-Zr; F6: Cr-Cu-Ni; F7: Br-I. Intensive effects are determined for the factor 1 distribution that is strongly correlated with dominant lithologic unit for whole territory of the Republic Macedonia (extracted as dominant lithogenic marker).
- The extracted geochemical association (including Cd-Pb-Zn-Sb) was defined as dominant anthropogenic marker for the atmospheric elements distributions. Significantly polluted sub-areas were determined, affected with significantly enrichments in Pb and Zn

atmospheric depositions and lower enrichments of Cd and Sb. “Zletovo” mine area and “Sasa” mine area were singled out as poly-metallic enriched areas; not only from the anthropogenic emission, but also the influence from the wind-blow dusting from soil surface was detected.

The historical records for elements deposition singled out the following conclusions:

- The activities carried out in the areas of poly-metallic hydro-thermal exploitation (*Sasa*, *Zletovo* and *Bučim* mines) lead to increased content of certain heavy metals in the atmosphere, which was determined through the monitoring of attic dust samples. The enforcement of statistical factor analysis singled out one dominant anthropogenic group of elements (Ag-Bi-Cd-Cu-In-Mn-Pb-Sb-Te-W-Zn). The fine particles from the mining operations are spread in the atmosphere carried by wind. Microparticles penetrate the human body through inhalation via the respiratory system.
- The dominant lithogenic markers that characterized the investigated area were extracted as following: Ga-Nb-Ta-Y-(La-Gd)-(Eu-Lu), Ga-Nb-Ta-Y-(La-Gd)-(Eu-Lu), Ba-Cs-Hf-Pd-Rb-Sr-Tl-Zr and K-Na-Sc-Ti. The geochemistry of the As-Co-Ge-V was also assumed as secondary affection from the anthropogenic emissions, also correlated with the dominance lithogenic occurrence of the Oligocene and Neogene volcanism.

Future work will allow refinement and more precise “fingerprinting” of geochemical processes, including the role of dust emissions and trace element-enriched accessory minerals in metasomatic events. Also, experiments are needed to extend the partition coefficient determinations to much greater assurance, to resolve controversies over the geochemical evolution.

REFERENCES

Aboal, J. R., Fernández J. A., Boquete, T., Carballeira, A. (2010). Is it possible to estimate atmospheric deposition of heavy metals by analysis of terrestrial mosses? *Science of the Total Environment,* 408(24), 6291-6297.

Aceto, M., Abollino, O., Conca, R., Malandrino, M., Mentasti, E., Sarzanini, C. (2003). The use of mosses as environmental pollution indicators. *Chemosphere*, 50, 333-342.

Agarwal, S. K. *(2009). Heavy metals pollution. A P H Publishing Corporation, New Delhi, India.*

Akimoto, H. (12003). Global Air Quality and Pollution. *Science*, 302, 1716-1719.

Alijagić, J., Šajn, R. (2014). Distribution of chemical elements in an old metallurgical area, Zenica (Bosnia and Herzegovina). *Geoderma*, 162, 71-85.

Alijagić, J., Šajn, R. (2006). Influence of ironworks on distribution of chemical elements in Bosnia and Herzegovina and Slovenia. *Geologija*, 49, 123-132.

Arsovski, M. (1997). *Tectonics of Macedonia*. Faculty of Mining and Geology, Štip, 1-306.

Artiolam J. F., Pepperm I., Brusseanm L. (2004). *Environmental monitoring and characterization*, Elevier Academic Press, San Diego.

Athar, M., Vohora, S. (1995). *Heavy metals and environment*, New Age International Publishers, New Delhi.

Bačeva, K., Stafilov, T., Šajn, R., Tănăselia, C. (2012). Moss biomonitoring of air pollution with heavy metals in the vicinity of a ferronickel smelter plant. Journal of Environmental Science and Health, Part A 47, 1-13.

Balabanova, B., Stafilov, T., Bačeva, K., Šajn, R. (2010). Biomonitoring of atmospheric pollution with heavy metals in the copper mine vicinity located near Radovıš, Republic of Macedonia. *Journal of Environmental Science and Health-Part A*, 45, 1504-1518.

Balabanova, B., Stafilov, T., Šajn, R., Bačeva, K. (2012). Characterisation of heavy metals in lichen species *Hypogymnia physodes* and *Evernia prunastri* due to biomonitoring of air pollution in the vicinity of copper mine. *International Journal of Environmental Research*, 6, 779-794.

Balabanova, B., Stafilov, T., Šajn, R., Bačeva, K. (2011). Distribution of chemical elements in attic dust as reflection of lithology and anthropogenic influence in the vicinity of copper mine and flotation. *Archives of Environmental Contamination and Toxicology*, 61, 173-184.

Balabanova, B., Stafilov, T., Šajn, R. (2015). Litogenic distribution of rare earth elements in automorphic and alluvial soils from Bregalnica River Basin. *Macedonian Journal of Chemistry and Chemical Engineering*, 34(1), 201-212.

Barandovski, L., Frontasyeva, M.V., Stafilov, T., Šajn, R., Ostrovnaya, T. (2015). Atmospheric deposition of trace elements in Macedonia studied by the moss biomonitoring technique. *Environmental Science and Pollution Research*, 22(20), 16077-16097.

Barandovski, L., Stafilov, T., Šajn, R., Frontasyeva, M.V., Bačeva, K., Air pollution study in Macedonia using a moss biomonitoring technique, ICP-AES and AAS. *Macedonian Journal of Chemistry and Chemical Engineering*, 32(1), 89-107.

Bourennane, H., Douay, F., Sterckeman, T., Villanneau, E., Ciesielski, H., King, D., Baize, D. (2010). Mapping of anthropogenic trace elements inputs in agricultural topsoil from Northern France using enrichment factors. *Geoderma*, 157(3), 165-174.

Carballeira, C. B., Aboal, J. R., Fernández, J. A., Carballeira, A. (2008). Comparison of the accumulation of elements in two terrestrial moss species. *Atmospheric Environment,* 42, 4904-4917.

Cekova, M. (2005). *Review of the bryoflora of the Republic of Macedonia.* Ss. Cyril and Methodius University, Skopje.

Cizdziel, J. V., Hodge, V. F. (2000). Attics as archives for house infiltrating pollutants: trace elements and pesticides in attic dust and soil from southern Nevada and Utah, *Microchemical Journal,* 64, 88-92.

Coşkun, M., Frontasyeva, M. V., Steinnes, E., Çotuk, Y., Pavlov, S.S. (2005). Atmospheric deposition of heavy metals in thrace studied by analysis of moss (*Hypnum cupressiforme*), *Bulletin of Environmental Contamination and Toxicology*, 74, 201-209.

Dumurdzanov, N., Serafimovski, T., Burchfiel, B.C. (2004). Evolution of the Neogene-Pleistocene Basins of Macedonia. *Geological Society of America, Digital Map and Chart Series 1.*

Fernández, J. A., Boquete, M. T., Carballeira, A., Aboal, J. R. (2015). A critical review of protocols for moss biomonitoring of atmospheric deposition: Sampling and sample preparation. *The Science of the Total Environment*, 517, 132-150.

Gosar, M., Šajn, R., Harald, B. (2006). Binding of mercury in soils and attic dust in the Idrija mercury mine area (Slovenia). *Science of the Total Environment*, 369, 150-162.

Greenwood, N. N., Earnshaw, A. (2005). *Chemistry of the Elements*, 2nd ed. Elsevier Butteworth-Heinemann, Oxford.

Harmens, H., Buse, A., Büker, P., Norris, D., Mills, G., Williams, B., Reynolds, B., Ashenden, T. W., Rühling, Å., Steinnes, E. ((2004). Heavy metal concentration in European mosses: 2000/2001 survey. *Journal of Atmospheric Chemistry*, 49, 425-436.

Harmens, H., Mills, G., Hayes, F., Norris, D. and the participants of the ICP Vegetation. (2013). *Air Pollution and Vegetation ICP Vegetation*, Annual Report 2012/2013 ICP Vegetation Programme Coordination Centre, Centre for Ecology and Hydrology, Environment Centre Wales, Bangor, UK.

Harmens, H., Mills, G., Hayes, F., Williams, P., De Temmerman, L. and the participants of ICP Vegetation Air pollution and vegetation.(2005). *ICP Vegetation Annual Report 2004/2005.* UNECE ICP Vegetation Coordination Centre, Centre for Ecology and Hydrology, Bangor, UK.

Harmens, H., Norris, D. A., Koerber, G. R., Buse, A., Steinnes, E., Rühling, Å. (2007). Temporal trends in the concentration of arsenic, chromium, copper, iron, nickel, vanadium and zinc in mosses across Europe between 1990 and 2000. *Atmospheric Environment*, 41, 6673-6687.

Harmens, H., Norris, D., Cooper, D., Hall, J. and the participants of the moss survey. (2008). *Spatial trends in nitrogen concentrations in mosses across Europe in 2005/2006.* Defra contract AQ0810. ICP Vegetation Programme Coordination Centre, Centre for Ecology and Hydrology, Bangor, UK.

Harmens, H., Norris, D. A., Sharps, K., Mills, G., Alber, R., Aleksiayenak, Y., Blum, O., Cucu-Man, S.-M., Dam, M., De Temmerman, L., Ene, A., Fernández, J. A., Martinez-Abaigar, J., Frontasyeva, M., Godzik, B., Jeran, Z., Lazo, P., Leblond, S., Liiv, S., Magnússon, S. H., Maňkovská, B., Phil-Karlsson, B., Piispanen, J., Poikolainen, J., Santamaria, J. M., Skudnik, M., Spiric, Z., Stafilov, T., Steinnes, E., Stihi, C., Suchara, I., Thöni, L., Todoran, L., Yurukova, L., Zechmeister, H. G. (2015). Heavy metal and nitrogen concentrations in mosses are declining across Europe whilst some "hotspots" remain in 2010. *Environmental Pollution* 151, 368-376.

Ilacqua, V., Freeman, N. C., Fagliano, J. P., Lioy, J. (2003). The historical record of air pollution as defined by attic dust. *Atmospheric Environment,* 37, 2379-2389.

ISO 14869-1:2001, Soil quality-Dissolution for the determination of total element content-Part 1: Dissolution with hydrofluoric and perchloric acids. International Organization for Standardization, Geneva, Switzerland.

Jeffrey, J., Davisa, B., Gulson, B. L. (2005). Ceiling (attic) dust: A "museum" of contamination and potential hazard. *Environmental Research,* 99, 177-194.

Jemec Auflič, M., Šajn, R. (2007). Geochemical research of soil and attic dust in Litija area, Slovenia. *Geologija*, 50, 497-505.

Kabata-Pendias, A. (2010). *Trace elements in soils and plants.* CRC press Boca Raton.

Kanaroglou, M., Jerrett, J., Morrison, B., Beckerman, M. A., Arain, N. L., Gilbert, J., Brook, R. (2005). Establishing an air pollution monitoring network for intra-urban population exposure assessment: A location-allocation approach. *Atmospheric Environment*, 39, 2399-2409.

Kaymaz, I. (2005). Application of kriging method to structural reliability problems. *Structural Safety*, 27, 133-151.

Keller, C. B., Schoene, B., Barboni, M., Samperton, K. M., Husson, J. M. (2015). Volcanic-plutonic parity and the differentiation of the continental crust. *Nature*, 523, 301-307.

Lazarevski, A. (1993). *Climate in Macedonia*, Kultura, Skopje.

Manara, A. (2012). Plant responses to heavy metal toxicity. In: A. Furini (Editor), *Plants and heavy metals*, Springer, London, UK.

Market, B. (2007). Definition principles for bioindication and biomonitoring of trace metals in the environment, *Journal of Trace Elements in Medicine and Biology*, 21, 77-82.

Marschner, H. (1995). Mineral nutrition of higher plants (2nd ed.). Academic Press Elsevier, London, UK.

Mitrev, S., Mihajlov. Lj., Micevski, T., Boev, B., Trajkova, F., Balabanova, B., Markova, N., Kletnikovski, C., Zlatkovski, V., Bikova, A., Golceva, Z., Tusevski, G., Sumanski, D. (2010). *Study on the sustainable development of the organic agricultural production in the Eastern planning region.* Goce Delcev, University, Stip.

Rakićević, T., Dumurdzanov, N., Petkovski, M. (1968). *Basic geological map of SFRJ, sheet Štip, M* 1:100,000 (map & interpreter), Federal Geological Survey, Beograd.

Rubatto, D. (2002). Zircon trace element geochemistry: partitioning with garnet and the link between U–Pb ages and metamorphism. *Chemical Geology*, 184(1), 123-138.

Rühling, Å., Tyler, G. (1973). Heavy metal deposition in Scandinavia. *Water, Air and Soil Pollution*, 2, 445-455.

Rühling, Å., Tyler, G. (1971). Regional differences in the heavy metal deposition over Scandinavia. *Journal of Applied Ecology*, 8, 497-507.

Šajn, R. (1999). Geochemical properties of urban sediments on the territory of Slovenia. *Geolological Survey of Slovenia*, Ljubljana.

Šajn, R. (2000). Influence of lithology and antropogenic activity on distribution of chemical elements in dweling dust, Slovenia. *Geologija*, 43, 85-101.

Šajn, R. (2001). Geochemical research of soil and attic dust in Celje area (Slovenia). *Geologija*, 44, 351-362.

Šajn, R. (2002). Influence of mining and metallurgy on chemical composition of soil and attic dust in Meža valley, Slovenia. *Geologija*, 45, 547-552.

Šajn, R. (2003). Distribution of chemical elements in attic dust and soil as reflection of lithology and anthropogenic influence in Slovenia. *Journal de Physique*, 107, 1173-1176.

Šajn, R. (2006). Factor analysis of soil and attic-dust to separatemining and metallurgy influence, Meza valley, Slovenia. *Mathematical Geology*, 38, 735-746.

Šajn, R. (2005). Using attic dust and soil for the separation of anthropogenicand geogenic elemental distributions in an old metallurgic area (Celje, Slovenia). *Geochemistry*, 5, 59-67.

Šajn, R., Žibret, G., Alijagić, J. (2012). Chemical composition of urban dust in Slovenia. In: L. B. Wouters &, M. Pauwels (Eds.), Dust: sources, environmental concerns, and control, (Environmental health - physical, chemical and biological factors), Public health in the 21st century. *Nova Science Publishers*, Hauppauge, NY.

Sakata, S., Ashida, F., Zako, M. (2005). Structural optimization using Kriging approximation. *Computer Methods in Applied Mechanics and Engineering*, 192, 923-939.

Sengupta, M. (1993). *Environmental impacts of mining: Monitoring, restoration and control*, Lewis Publishers, Boca Raton.

Serafimovski, T., Aleksandrov, M. (1995). Lead and zinc deposits and occurrences in the Republic of Macedonia. Special edition of the Faculty of Mining and Geology, Stip, Macedonia.

Sokhi, R. S. (2011). *World atlas of atmospheric pollution*. Anthem press, London, UK.

Stafilov, T., Bojkovska, R., Hirao, M. (2003). Air pollution monitoring system in the Republic of Macedonia. *Journal of Environment and Protection Ecology*, 4, 518–524.

Stafilov, T., Balabanova, B., Šajn, R., Bačeva, K., Boev, B. (2010). *Geochemical atlas of Radoviš and the environs and the distribution of heavy metals in the air*. Faculty of Natural Sciences and Mathematics, Skopje.

Stafilov, T., Šajn, R., Pančevski, Z., Boev, B., Frontasyeva, M. V, Strelkova, L. P. (2008a). *Geochemical atlas of Veles and the environs*. Faculty of Natural Sciences and Mathematics, Skopje.

Stafilov, T., Šajn, R., Boev, B., Cvetković, J., Mukaetov, D. (2008b). Andreevski M. (2008b). *Geochemical atlas of Kavadarci and the environs*. Faculty of Natural Sciences and Mathematics, Skopje.

Stafilov, T., Šajn, R., Sulejmani, F., Bačeva, K. (2011). *Geochemical atlas of Kičevo and the environs*. Faculty of Natural Sciences and Mathematics, Skopje.

Tyler, G. (2008). Bryophyte and heavy metals: A literature review. *Botanical Journal of the Linnean Society,* 104, 231-253.

Vallero, D. (2008). Fundaments of air pollution, 4th edition, Elsevier Academic press, London, UK.

Van het Bolcher, M., Van der Gon, D. H., Groenenberg, B. J., Ilyin, I., Reinds, G. J., Slootweg, J., Travnikov, O., Visschedijk, A., de Vries, W. (2006). *Heavy Metal Emissions, Depositions, Critical Loads and Exceedances in Europe*. Hettelingh J P. & Sliggers J. (eds.), National Institute for Public Health and the Environment, The Netherland.

VanLoon, G. W., Duffy S. J. (2000). *Environmental Chemistry: A global perspective*, Oxford University press, New York.

Wang. L. K., Chen, J. P., Hung, Y. T., Shammas, N. K. (2009). *Heavy metals in the Environment*. CRC Press Taylor & Francis Group, Boca Raton, FL.

White, W. M. (2013). *Geochemistry*. Wiley-Blackwell, Chichester.

Žibret, G., Šajn R. (2008a). Impacts of the mining and smelting activities to the environment - Slovenian case studies. In: M. L. Sánchez (Ed.), Causes and effects of heavy metal pollution. *Nova Science Publishers*, New York.

Žibret, G., Šajn, R. (2008). Modelling of atmospheric dispersion of heavy metals in the Celje area, Slovenia. *Journal of geochemical exploration*, 97, 29-41.

Žibret,_G., Šajn, R. (2010). Hunting for geochemical associations of elements: factor analysis and self-organizing maps. *Mathematical Geosciences*, 42, 681-703.

Žibret, G. (2012). Impact of dust filter installation in ironworks and construction on brownfield area on the toxic metal concentration in street and house dust (Celje, Slovenia). *Ambio*, 41, 292-301.

RELATED NOVA PUBLICATIONS

ENVIRONMENTAL AND REGIONAL AIR POLLUTION

Dean Gallo and Richard Mancini

ISBN: 978-1-60692-893-6
Publication Date: 2009

The study of the environment includes interactions among the physical, chemical and biological components of the environment. The introduction of contaminants into the environment that cause instability, disorder, harm or discomfort to the physical systems or living organisms is a major concern that is addressed in this book. A simple field sampling/preconcentration strategy for determination and speciation of Cr(III) and Cr(VI) in seawater is described. Advantages of green methodologies over traditional techniques to extract pesticides from the environment are also presented. Also included in this book is a review of the most important sources of Hg into the environment and an evaluation of the use of different animal species as bioindicators of human mercury exposure. A variety of regional air pollution issues including the toxicity of smoke and the harmful effects of high surface ozone concentrations effects on human's health and crop production are examined.

Traffic Related Air Pollution and Internal Combustion Engines

Sergey Demidov and Jacques Bonnet

ISBN: 978-1-60741-145-1
Publication Date: 2009

Atmospheric pollution has been a major problem in human technological development and motor vehicles are one of the major sources of particulate matter pollution. This book investigates current models designed to predict air pollutant emissions and fuel consumption for road traffic and presents the outputs of statistical models developed to derive emission factors. Information on the use of Geographic Information Systems and traffic area air pollution monitoring stations is presented in order to comprehend the variations of traffic-related air pollution. Furthermore, this book reports the pros and cons of hydrogen-fueled internal combustion engines, a study of the new technology to produce syngas from methane with a compression ignition engine. An overview of the characteristics of the factors influencing the thermal efficiency of spark ignition engines fueled with hydrogen is given as well.

Air Pollution and Ship Emissions

Jacob Boutin

ISBN: 978-1-60876-087-9
Publication Date: 2010

As pollution from cars, trucks, and land-based stationary sources has been more tightly controlled over the last 40 years, the contribution of ships and port operations to air pollution in port cities has become more important. In the same period, foreign trade has grown dramatically; thus, pollution from shipping and port operations would be growing as a percentage of total emissions, even if the emissions were regulated to the same degree as other sectors. This book provides information regarding pollution from ships and port facilities and discusses some of the measures being implemented and considered by local, state, and federal regulatory agencies. This book consists of public documents which have been located, gathered, combined, reformatted, and enhanced with a subject index, selectively edited and bound to provide easy access.

ENCYCLOPEDIA OF AIR POLLUTION (2 VOLUME SET)

Mark E. Sheffield

ISBN: 978-1-61761-833-8
Publication Date: 2013

This book presents important research advances in the study of air pollution, as well as traffic-related air pollution, the national air quality and the role of ship emissions in air pollution. Topics discussed herein include traffic benzene pollution in Bangkok; organic pollutants in urban air; development of forecasting models for predicting hourly ozone concentrations; traffic-related air pollution; snowmobile pollution in North America; hydrogen-fueled internal combustion engines; air quality standards; air pollution and greenhouse gas emissions from ships; and emission control area proposals.

AIR POLLUTION EMISSIONS

Daniel G. Vasiliev and Robert A. Kask

ISBN: 978-1-62100-453-0
Publication Date: 2012

In this book, the authors present topical research in the study of air pollution emissions. Topics discussed include opportunities for sustainable energy development and applications in building environments and air pollution controls; hydrocarbon air pollution laser monitoring; inventory and behaviour of air pollutants in an urban area of NW Spain and spatial modeling of air pollution based on traffic emissions.

AIR POLLUTION: SOURCES, PREVENTION AND HEALTH EFFECTS

Rajat Sethi (Texas A&M Health Science Center (TAMHSC), Bryan, Texas, USA)

ISBN: 978-1-62417-735-4
Publication Date: 2013

Industrial pollution is the largest fountain of air offenses in most of the developed world. Statistics from the U.S. Environmental Agency confirm that more than half of the nation's total pollution is courtesy of industry. What's more, six out of every ten people in the United States live in urban areas and are currently inhaling these bi-products. The consequence of air misuse is glaring: approximately four percent of all deaths in the United States are attributed to air pollution. This book highlights the sources and effects of air pollution, and also puts forth solutions. Successful pollution prevention is essential to our future well-being. Quintessential to this end, is prevention through education; the goal of this publication.

INDEX

C

D

E

M

N

O

P

R

S

T

U

V

W

Z